Building the Trident Network

Inside Technology
edited by Wiebe E. Bijker, W. Bernard Carlson, and Trevor Pinch

Janet Abbate, *Inventing the Internet*

Charles Bazerman, *The Languages of Edison's Light*

Marc Berg, *Rationalizing Medical Work: Decision-Support Techniques and Medical Practices*

Wiebe E. Bijker, *Of Bicycles, Bakelites, and Bulbs: Toward a Theory of Sociotechnical Change*

Wiebe E. Bijker and John Law, editors, *Shaping Technology/Building Society: Studies in Sociotechnical Change*

Stuart S. Blume, *Insight and Industry: On the Dynamics of Technological Change in Medicine*

Geoffrey C. Bowker, *Science on the Run: Information Management and Industrial Geophysics at Schlumberger, 1920–1940*

Geoffrey C. Bowker and Susan Leigh Star, *Sorting Things Out: Classification and Its Consequences*

Louis L. Bucciarelli, *Designing Engineers*

H. M. Collins, *Artificial Experts: Social Knowledge and Intelligent Machines*

Paul N. Edwards, *The Closed World: Computers and the Politics of Discourse in Cold War America*

Herbert Gottweis, *Governing Molecules: The Discursive Politics of Genetic Engineering in Europe and the United States*

Gabrielle Hecht, *The Radiance of France: Nuclear Power and National Identity after World War II*

Kathryn Henderson, *On Line and On Paper: Visual Representations, Visual Culture, and Computer Graphics in Design Engineering*

Eda Kranakis, *Constructing a Bridge: An Exploration of Engineering Culture, Design, and Research in Nineteenth-Century France and America*

Pamela E. Mack, *Viewing the Earth: The Social Construction of the Landsat Satellite System*

Donald MacKenzie, *Inventing Accuracy: A Historical Sociology of Nuclear Missile Guidance*

Donald MacKenzie, *Knowing Machines: Essays on Technical Change*

Donald MacKenzie, *Mechanizing Proof: Computing, Risk, and Trust*

Maggie Mort, *Building the Trident Network: A Study of the Enrollment of People, Knowledge, and Machines*

Susanne K. Schmidt and Raymund Werle, *Coordinating Technology: Studies in the International Standardization of Telecommunications*

Building the Trident Network
A Study of the Enrollment of People, Knowledge, and Machines

Maggie Mort

The MIT Press
Cambridge, Massachusetts
London, England

© 2002 Massachusetts Institute of Technology

All rights reserved. No part of this book may be reproduced in any form by any electronic or mechanical means (including photocopying, recording, or information storage and retrieval) without permission in writing from the publisher.

Set in New Baskerville by The MIT Press.
Printed and bound in the United States of America.

Library of Congress Cataloging-in-Publication Data

Mort, Maggie.
Building the Trident network : a study of the enrollment of people, knowledge, and machines / Maggie Mort.
p. cm.
Includes bibliographical references and index.
ISBN 0-262-13397-0 (hc. : alk. paper)
1. Trident (Weapons systems)—Design and construction. 2. Military-industrial complex—United States. 3. United States. Navy—Procurement. 4. Military-industrial complex—Great Britain. 5. Great Britain. Royal Navy—Procurement. I. Title.
V993 .M67 2001
338.4'762382574—dc21   2001034275

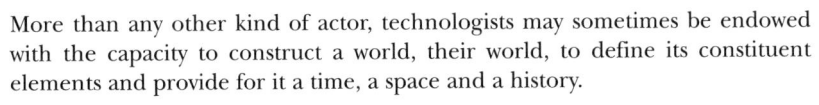

More than any other kind of actor, technologists may sometimes be endowed with the capacity to construct a world, their world, to define its constituent elements and provide for it a time, a space and a history.
—*Michel Callon, 1986*

Contents

Acknowledgments ix

Introduction 1

I
Simplifying Production

1
Large-Scale Technologies: Ambivalence and Coercion 17

2
From Diversity to "Core Business" 33

II
Reinforcing the Network

3
Technological Roads Not Taken: The Constant Speed Generator Drive 53

4
Constructing a Core Workforce 79

III
Alternatives

5
Winning the "Technical Arguments" 111

6
Building a Counter-Network 129

IV
Closures

7
Human Redundancy: An Exercise in Disenrollment 161

8
Softening the Facts 175

Appendix Employment Data from Lazard Brothers Sale Documents, as Reproduced in Oceans of Work 185

Notes 187
Bibliography 205
Index 213

Acknowledgments

This book is dedicated to Danny Pearson, former shipyard union convenor, who died suddenly shortly after I began my research at Barrow. He cared passionately about the effects of unemployment and wastage of skill in the shipyard and its community. Though he had retired, his enthusiasm for diversity and workplace creativity was undiminished, and it prompted him to offer me any assistance I needed to recover the history of the 1980s shipyard alternatives campaign. Unfortunately, our acquaintance was very short, and later there were times when I might have lost the determination to continue. I feel that this book would not have been completed without the spur that Danny's unshakable conviction continued to provide.

I want to thank all those people who granted me interviews in the course of this study. Some could not be named in the body of the text for reasons of confidentiality, yet many of these anonymous interviewees gave me substantial help and information. The former members of the Barrow Alternative Employment Committee gave their time unstintingly to my attempts to recover memories, dates, and documents. In particular I would like to acknowledge my debt to Betty Pearson, Terry McSorley, Bob Bolton, Steven Schofield, and Alan Milburn. Also Keith Sutton (for use of the *Evening Mail* library), Peter Murrell, Hilary Wainwright, and Bernard Harbor for early encouragement. My discussions with Donald MacKenzie and Graham Spinardi were invaluable; without them the book's limitations would be considerably greater. I owe special thanks to Hilary Corton and to Mike Michael.

Two personal experiences strongly influenced (and admittedly limited) the context for this study. Before taking up an SERC/ESRC studentship at Lancaster University, I worked as a reporter on the *North West Evening Mail* in Barrow. During my time with the newspaper, it was the nature of the relationship between the town, the shipyard, and the

Trident contracts that interested me, though there was never time to explore this. When I resigned to take up the studentship and was introduced to the sociology of scientific knowledge, I realized that it might be possible to return to these issues equipped with new insights.

On two occasions during my 6 years as a journalist, it fell to me, as a workplace representative of the National Union of Journalists, to undertake detailed and lengthy negotiations, alongside other trade union colleagues, over the introduction of "new technology" in newspapers. The issue of "direct input" in newspapers had been hugely controversial since the bitter Wapping dispute between Rupert Murdoch and the trade unions. The NUJ was facing derecognition, and the print union, then the National Graphical Association, was facing obliteration. The new technology agreements that we reached in two very different workplaces were the result of a process of simultaneous engagement with technology and politics and with issues such as retraining, ergonomics, industrial action, and redundancies. The technology was the stage on which it all happened. Thinking back, I realize that we all had to be "heterogeneous engineers"—we had to make our claim to be full actors in the process of change, rather than peripheral ones.

Building the Trident Network

Introduction

On Saturday, September 19, 1998, the last Trident submarine emerged into bright sunshine from its massive construction shed at the Barrow in Furness shipyard.[1] I watched from Michaelson Bridge opposite the yard, along with a scattering of local people and a few hundred anti-nuclear demonstrators. The ceremony for this fourth and final vessel in the most costly weapons program ever undertaken in the United Kingdom was low-key. The project was now something of an embarrassment to the recently elected Labour government. Already people were forgetting the Cold War, for which the Trident program was intended. Press and media coverage was virtually nil. The story was over, the world had moved on.

The few local spectators watched gloomily. Nuclear-powered-and-armed submarines are not objects of celebration, for all the skill and achievement they embody. They are black, silent creatures, designed to disappear with their deadly cargoes, thereafter to remain unseen by the general citizenry. Townspeople also feared that the launch might add to the stream of job losses already inflicted as the shipyard moved through the program's four contracts.

As ever, the locals tolerated the demonstration, pathetically small this time compared with the torrents of protesters seen in Trident's early years. To add to the sense of redundancy, as we turned away from the launch and looked down the Walney Channel, we saw the gleaming outlines of the four smaller Upholder Class submarines that were completed during the early stages of the Trident program but never brought into use.

This book traces the path that led to the final launch. It is the story of a large-scale technical system, of that system's impact on smaller systems, and, most important, of that system's impact on the local communities of people engaged in its production. It is a story of courage, heartbreak, betrayal, and humor, and of human and technical achievement coexisting with waste and destruction. What forces, what details, and what nurturing kept the colossal Trident program moving forward?

The Special Relationship: Controversy and Contingency

The new generation of nuclear submarines was proving so complex, so sophisticated as to rival in scope the great medieval cathedrals of Europe, where tradesmen passed their tasks from generation to generation, each hoping he would be among those to pray inside.
—*Tyler 1986, p. 323*

This statement reflects the experience of the United States in constructing a nuclear submarine fleet. The US program of the 1970s and the 1980s was a vast undertaking: at any one time, the premier US shipyard might have been working on five or six nuclear submarines, each containing more than 80,000 separate welds (Tyler 1986, pp. 54 and 261).

The United Kingdom's program to construct four new-class submarines to provide a platform for the Trident weapons system was much smaller, but it was still the largest and most ambitious technological enterprise ever undertaken by the UK. This program was, within its own terms, a success. Even when the large-scale "weapons succession" process that characterized the Cold War era was largely halted after the collapse of the Soviet bloc, the construction of the Trident system moved ahead, seemingly without encountering either technical or political problems.

Vickers Shipbuilding and Engineering Ltd.[2] (VSEL) performed a remarkable feat of engineering and project management in producing *Vanguard*, the UK's first Trident submarine, largely on schedule and within budget. For comparison, consider that the Electric Boat Company, owned by the industrial giant General Dynamics, the world's largest submarine builder, delivered the first ('lead') US Trident submarine, the *Ohio*, 3 years late and 55 percent over the contract's ceiling price (Schumacher 1988, p. 1).[3] The *Vanguard* and the *Ohio* were produced under different conditions, of course, but this difference in production performance on the lead submarine does suggest the high achievement of VSEL's workforce.

The purpose of this study is to find out how the Trident system[4] became accepted, stabilized, produced, and "black boxed" in the UK, in such a way that it has now taken on the appearance of having fulfilled some kind of "technological trajectory." Where were its contingencies, weaknesses, and controversies, and how were these managed and overcome?

When, in 1981, the United Kingdom decided to follow the lead of the United States and equip its nuclear submarines with the enhanced D5 version of the Trident missile, it was opting for a huge increase in strategic nuclear capability. Just a year earlier, a decision to buy the lesser Trident C4 had been taken in secret by a small cabinet committee and only later announced to the full cabinet and to Parliament. Both decisions were arrived at in a similar way, and both immediately aroused widespread controversy (Ponting 1989, pp. 177–191). Both the Labour opposition and the Liberal Party committed themselves to cancellation of the new program. The decisions were also fiercely opposed by antinuclear groups, which at the time were experiencing a surge in membership and a period of dynamic activism. The high cost[5] of Trident also alienated certain defense and naval chiefs whose own programs seemed threatened by expansion of the role of submarine-launched ballistic missiles. There was even concern among back-bench Conservative Members of Parliament (MPs) about the displacement effect Trident would have on the overall naval budget, with consequent job losses in their constituencies. Moreover, the decision to "buy American" once again (as had been done in the Polaris program) was not popular with many sections of the UK's defense industry, which had to be content with eventual concessions in the sales agreement that allowed UK firms to bid (largely unsuccessfully, as it turned out) for Trident supply contracts. Buying American was also unpopular with those MPs, mainly from the political right, who wanted the UK's nuclear "deterrent" to be more independent and would have preferred to see the UK develop its own program. Finally there was opposition from a strand of trade-union opinion that saw defense dependence, particularly on increasingly large-scale systems vulnerable to cancellation, as a high-risk employment strategy.[6] And some workers were uneasy about their role in an increasingly offensive, rather than defensive, arms race.

The controversy in the United Kingdom contrasted markedly with the situation in the United States, where political and labor controversy was relatively insignificant. Using a social-constructionist approach to examine the "technical"[7] problems encountered by the US Trident builders before the system was black boxed, Spinardi (1994) and MacKenzie (1990) were able to show that the Trident system's technical characteristics and components were products of both social and technical processes. Their approach, which views Trident as a "sociotechnical network," is extremely helpful for the study of the system's stabilization in the United Kingdom, where access to official information is more problematic than in the United States and access to unofficial information is hampered by secrecy and self-censorship.

The stabilization of the US Trident program involved the manipulation of congressional committees, budgets, and artifacts in a process that has been called "heterogeneous engineering."[8] Public controversy (more specific, workers' opposition to the program) does not seem to have been a major issue.[9] In the US, "heterogeneous engineers" progressively ironed out many sociotechnical difficulties in constructing the system; the UK "engineers"[10] encountered their share of technical problems, but in a context of delicate and volatile political uncertainties. How did these engineers overcome the peculiar political and technical problems that faced Trident?

This book is situated somewhere between the literature on arms control and disarmament and that of science and technology studies (STS). My point of view is that of labor; thus, I refer at various points to the literature of resistance and to alternatives that emanate from workplace knowledge. It is this perspective that I argue connects the disarmament debate with that of STS. This point is illustrated by the decision to adopt the US-enhanced Trident missiles. "Buying American" was seen as the lowest-cost option for the UK to maintain strategic capability (the alternative being a risky indigenous development). When it came to choosing a submarine design and propulsion system, though, the Thatcher government chose the most advanced, expensive, and futuristic platform option, designed and made in the UK, with all the technical risks attendant on utilizing untried technologies. By March 1982 it was decided that an entirely new generation of submarine, the Vanguard Class, would be designed and constructed, and that it would incorporate part of the American Ohio Class design to accommodate a midsection missile compartment. The Vanguard Class would also utilize the latest pressurized-water reactor, a PWR-2, for propulsion; this reactor was still under development at the Dounreay plant in northern Scotland. An enlarged submarine (though not necessarily a whole new design) was essential to accommodate the very latest missile, the Trident D5. With the emergence and adoption of the enhanced D5, the United States had transformed the rhetorical purpose of its Trident system from its earlier deterrent role (an integral part of the strategy of mutually assured destruction) into a deliverer of "counterforce," a defense logic in which nuclear war might actually be winnable through the achievement of "hard-target kill."[11] The UK's decision to abandon the C4 in favor of buying the D5 showed the extent to which the UK had become tied into both US strategy and US production of the missile. The C4 had been the UK's preferred option, and the extra, escalated destructive capability and political consequences of adopting the D5 system (along with the technical risks outlined above) came to be justified by the logic of maintaining "commonality" with the US. "Commonality," in this context, was a term that encompassed the notion that technical and political goals were inseparable.

McInnes (1986) has pointed out that, once the US adopted the D5 missile, it became clear that existing C4 production lines would be closed down earlier than had been planned. The UK's options were thus restricted: the missiles would have had to be bought earlier than the otherwise might have been, and the cost of maintaining the C4 unilaterally would have been incurred. The alternative was to take on a weapons-system capability that the government had previously stated it did not need, claiming economic and operational necessity as the reason (ibid., p. 51).[12] Though strategic and operational risks were said to be minimized by the decision, this was at the expense of political risks and technical uncertainties involved in accommodating the new missile.

Radical or Conservative Technology?

An STS approach will allow a richer analysis of the construction of the Trident system than is usually found in the literature of arms control. For example, an STS perspective draws our attention to the fact that Trident development has been marked by *interpretive flexibility*: at different stages of development in both the United States and the United Kingdom, the system was variously described as technology that was radical and challenging and as technology that was traditional and utilized existing skills.

In the United States, the lead submarine, the *Ohio*, was the subject of an estimated 100,000 drawings and was described by Commander Edward Peebles, US deputy program manager for Trident Ship Acquisition, as "probably the biggest mobile structure built by man." Admiral Robert Gooding, Commander of Naval Ship Systems, believed that, with their new propulsion and weaponry advances, Trident submarines would be strikingly different from Polaris submarines. Both men were supporters of cost-reimbursement (cost-plus) contracts under which the US government would bear much of the development costs. In contrast, Vice-Admiral Hyman Rickover, head of Naval Reactors Branch, was supported by Admiral Harvey Lyon, Trident Program Manager, in maintaining that the shipbuilders had "built Polaris submarines, and Trident is not that much different" (Schumacher 1988b). Both of these men supported fixed-price contracts under which the US Navy would, they believed, retain more control over the shipbuilder. The argument seems to have crystallized over the proposed escalation of the program by President Nixon in order to put pressure on the Soviet Union in the Strategic Arms Limitation Talks (Schumacher 1987, 1988b).

In the United Kingdom, Trident was also variously represented as somehow both conservative and radical. Early in 1993, Lord Chalfont, the chairman of VSEL, said that VSEL's design and construction workers were "now capable of producing a piece of engineering like the Vanguard, the Trident submarine, which in putting it together is something like putting a man on the moon in terms of advanced concepts and engineering." This quote is from an interview with Julian O'Halloran for a segment of the BBC1 television program *Panorama* titled "The Peace Penalty: Whatever Happened to the Peace Dividend?" (broadcast March 22, 1993). Chalfont was at that time making a plea for more defense contracts to be placed with VSEL. In contrast, defending the government's non-interventionist policy in the face of mass redundancies[13] at the shipyard, Michael Heseltine, president of the Board of Trade, in an interview for the same TV program, said that the UK could not go on "supporting skilled teams in existence doing the things they were doing yesterday, serving yesterday's markets." When I interviewed him on June 28, 1993, Eric Montgomery, a former trade-union convenor[14] in the UK's Trident shipyard who was campaigning for industrial conversion away from defense, described Trident as "just another boat." Trident could be radical or conservative, depending on what social and/or political objective was being sought. Mary Kaldor explores this point further in *The Baroque Arsenal* (1981), describing the creation of large-scale weapons technology as a contradictory process in which cutting-edge technical development is embodied in conservative products.

Recovering Contingency

In the history of the UK's Trident program, it is the submarine that brings together the components of the whole Trident system. Similarly, it is mainly the submarine builders who informed my study, in which a few of the possible histories of producing a huge sociotechnical system are told. However, these stories were not discovered or chosen randomly. Because this study concerned the construction of the UK's primary strategic nuclear weapons platform, there was a strong motivation in the selection of material to present episodes that would show where that technology was contingent or unstable. Though all technologies are, I believe, contingent at some or all stages in their history, when it comes to nuclear weapons there is a strong "you can't uninvent the bomb" rhetoric overlaying their development. From this rhetoric it follows that building the four Trident submarines was an inevitable, natural consequence of developing the missile. Drawing on insights from the sociology of scientific knowledge and from social-constructivist studies of technology, and employing an actor-network approach, I hope to show that the production of Trident was *not* inevitable.[15] My methodology centers on re-opening controversies that occurred during the planning and construction of Trident in the UK. One problem is that there are now few visible signs of controversy; indeed, the building of Trident is now presented as a great success.[16] Also, even if the study were confined to the construction of the UK's Trident *submarines*, this still remained a vast and complex network in which problems and controversies ranged from the physical (such as finding adequate boiler-shop space at the shipyard) to the social-psychological (e.g., maintaining the Cold War rationale for the system).

The most visible of the controversies concern Trident's political unpopularity in the United Kingdom in the early 1980s.[17] "The 1980s," Kaldor commented at the time (1981, p. 8), "may turn out to be one of those rare moments in history when real change is possible. Modern military technology is currently in crisis. This is manifested in the 'unreadiness' of the armed forces, the financial problems of the armorers, and above all, the growing disaffection of soldiers and defense workers in many countries." For builders of weapons systems and for the Thatcher government, the 1980s were politically uncertain years.[18] But what was the relationship between the national political controversy over adopting the system and possible controversies over technical choice at the point of production—the macropolitics and the micropolitics, the local and global networks? As I will explore in later chapters, the production of Trident involved savage rationalization and homogenization of skills and resources and rejection of technical "alternatives." If I were to think of Trident as a sociotechnical network (because the local production of the submarine is embedded in a national and a global network), I could not treat national "political" controversy and local "technical" controversy as separate issues.

Early in my research I discovered that a group of workers had mounted a campaign around the shipyard to promote alternatives to producing Trident. Their place in the history of Trident is almost unknown, but their activities, outlined in part III of the book, raised crucial questions about the construction of Trident. This group, the Barrow Alternative Employment Committee (BAEC), forged its own particular social and technical network that operated within (and because of) the emerging Trident network. In campaigning for alternative technologies, the group was in effect acting to dissolve the barrier that gets created between the technical (here standing for local production) and the socio-political (here standing for macropolitics). As shop stewards at Lucas Aerospace had recently discovered, this barrier also stood between workers, who were expected to campaign over wages and working conditions only, and management (sometimes supported by official national trade unions), which was seen as the legitimate decision maker and arbiter of the content of production.

The campaign by the BAEC brought the national political uncertainties and controversies about the Trident system together with the local problems about technical content, diversity, and employment at the point of production. The Barrow shipyard had been occupied with the design and construction of a series of Trafalgar Class hunter-killer submarines, but they were nearing completion and no other submarine program was envisaged. The members of the committee were living at both the ethical and the economic "sharp end" of the emerging Trident network: as they often pointed out, they and their community would inherit many of the social consequences of the company's exclusive dependence on Trident production.

I decided to follow this group of actors by reconstructing their history. I will refer to the BAEC and to its principal report (titled Oceans of Work) throughout this study. Because this group acted to unite the technical and the social, I have to some extent treated all the episodes recounted here as constituting the context for Oceans of Work. This is not, however, a case study of BAEC or even of Trident production itself; it is a study of the surrounding events. It is a story about how Trident production was organized and ordered. The story can never be complete; as John Law (1994, p. 45) points out, there are "discontinuities in ordering."

Frequent reference to the BAEC and to its report provided a way of creating a narrative about Trident submarine production and about the history of VSEL (the key design/construction company) that did not focus on the views and actions of managers. I have gone about this first by identifying a network within a network (the BAEC) and attempting to describe the relationship between the two and second by attempting to theorize the relationship between the dominant (Trident) network and its marginalized or expelled actants: workers who had been made redundant and technologies that had been abandoned. (This latter point is developed in chapter 7.) If prejudice, preconception, and social ordering usually prohibit the equal treatment of all actors in narratives about technology (Latour 1987, p. 258),[19] what I try to do here is to even their dimensions up a bit. If in this book the reader can approach the story of the building of Trident by first listening to participants in that network who were marginalized, or whose stories have had to be positively unearthed, the reader can (heuristically) bracket off those "power relations" that led to Trident's "success." In this way, it is possible to explore the "how"—the processes by which power relations were configured and eventually rendered fixed and "inevitable"—by asking questions as if power is something that circulates rather than being fixed.[20] Thus, the investigation takes place in a time before roles and power were distributed, before "facts and machines are blackboxed" (Latour 1987, p. 258). We are then able to see the ordering, not the order (Law 1994, p. 26).

One consequence of retracing the footsteps of these actors is that we are able to "forget" (or postpone the knowledge) that they were later marginalized, or that others later came to be pivotal. We can become immersed in their story, in their activities, almost as they did. By re-opening controversies and searching for alternative accounts, we can see how power inequalities were constructed and maintained, how power relations had to be policed, and how divisions between the technical and the social had to be continually reinforced and expanded.

My allegiance to an actor-network approach survived one important hurdle. Terry McSorley, a former shipyard worker and a founder of the BAEC, read chapters as they were being written. Commenting on the part about treating actors (in this instance the committee and the shipyard management) symmetrically and as equals in order to explore the circulation and potential accessibility of power, he said (referring to the activists' approach to the company): "Treating all the parties as equal—that's just what we did on the committee!" On the face of it, there was hardly any chance that a small committee of workers, struggling for acceptance even among their own trade-union organizations, would be able to alter the direction of production at the UK's premier nuclear submarine shipyard. But in spite of its apparent "powerlessness" (in actor-network terms, the apparent poverty of its associations with key players), the committee carried on, mustering resources to promote its alternative production strategy and its call for the content and the context of technology to be considered together in the interest of maintaining employment.

My study of the BAEC relies mainly on in-depth interviews with its most active members. The committee was certainly not homogeneous. Each member had a somewhat different view of the campaign,[21] and the committee also reflected the various trades and skill groups in the shipyard and the different stances of their unions. But two issues were common to all accounts of the strategies used. It was agreed first that the campaign should concentrate on protecting jobs, and second that all the campaigners should be active VSEL workers. By exploring and questioning the technical content of production, the committee members transcended sharp differences between trades, between unions, and between white-collar and manual sections, and they therefore formed a unique political alliance in the shipyard.[22]

Following the actors has also implied a degree of "snowballing"—picking up other interviewees from existing contacts. The snowballing also extended to the stories. For example, one of the issues raised by the BAEC in the 1980s was VSEL's failure to develop and exploit the Constant Speed Generator Drive (CSGD). This set me off: Who were the actors in the development of the CSGD? A trip to the Patent Office revealed the names of the inventors. A journey to Wales unearthed a rich story about the evolution of the technology and tensions between the English and Welsh production sites and also the (then nationalized) parent company, British Shipbuilders. This tension concerned the direction of technical development and the conflicting sociotechnical demands of the civil and defense markets. This story, told in chapter 3, was the sort of road not taken that, as David Noble (1984, p. 146) has pointed out, can help us understand the society that denied it.

All this prompted an attempt to amass all the available literature on the company (VSEL and its forerunner, Vickers) and its technological history. The rich diversity of the company's products, repeatedly mentioned by former managers and workers in interview, was only partially reflected in the official company histories. Other forms of literature, particularly sales brochures that emanated from the former engineering division, revealed just how important a feature of the Barrow-based business this diversity had been. It became important to try to understand this discrepancy between official and unofficial histories. National politics (nationalizations and denationalizations), in addition to pressure from the Ministry of Defence (MoD) to bring about single-purpose production, are part of the multi-stranded impetus for the disappearance of general engineering at the shipyard. Non-submarine work, though often profitable, came to be seen as a distraction from "core business" (that is, large MoD contracts). But the local engineers, proud of in-house skills and design achievements, talked freely about the former diversity of their work. This construction of "core business" and the associated shrinkage of products is discussed in chapters 2 and 3.

Shipyard workers sometimes signaled their ambivalence toward building Trident, an unease that was picked up in local street surveys conducted by anti-nuclear groups in the 1980s. I tried to uncover more about this ambivalence, looking at how workers were persuaded to see Trident as aligning with their own interests. This led to a study of the 1986 privatization and the company's share-option scheme, an exercise in local "enrollment" of employees into a new experience of capitalism and into tighter production goals.

My extensive use of cuttings from the *North West Evening Mail* should be explained. Because the community of Barrow and the Furness peninsula of Cumbria as a whole are geographically isolated, the local newspaper played a significant role in both reflecting and influencing local opinion. Like it or loathe it, most local people read this daily tabloid, and its sales have always relied considerably on VSEL workers' buying it at the factory gate. In fact, in 1990, when VSEL increased its shift hours between Mondays and Thursdays, allowing workers to finish early on Fridays, the *Mail* altered its production schedule to bring out the Friday edition in time for the new midday shipyard exodus.

The *Mail*'s editorial policy was ambivalent. It supported VSEL and Trident, both because they provided jobs and because they had made Barrow a major player in maintaining the national defense.[23] Yet the *Mail* also ran many prominent features about the devastation caused by the redundancy program of the early 1990s, when workers were leaving the town to find work.[24] The *Mail* also tried to shame VSEL into justifying the profits it amassed while thousands of workers were being laid off. Because it was widely read by workers and their families, VSEL's management (and to a lesser extent the trade unions) attempted to use the *Mail* as a mouthpiece. The *Mail* was therefore a rich source of information for an investigation into how workers were persuaded to see their interests as aligning with those of the company and the Trident contracts. In addition, the *Mail* could not ignore the effects of what I have called "disenrollment"—the jettisoning of people and machines from the network.

The concept of technical and human disenrollment from networks is developed in later chapters, which consider how both people and machines can be rendered redundant. Science and technology studies have tended toward the heroic, examining how new technologies are built, how human and nonhuman participants become engaged, and how the network is finally stabilized. But what of networks in decline or reduction? What of the actors being expelled from the network? I have developed the concept of disenrollment to describe the cruel functionality of redundancy, and to examine how this can reinforce a network and assist in the stabilization of what remains. Many of the individuals I interviewed reflected on redundancy and its effect on individuals and the community. Many became angry when they talked about the company's method of implementing redundancy and its formalized attempt, by giving each employee a numerical score, to make redundancy appear inevitable and immutable. And this construction of a sense of inevitability about redundancy led me back to the initial questions about the aura of inevitability that came to surround Trident itself.

Organization of the Book

Chapter 1 reflects on how large-scale technologies have been theorized and how some of these approaches can be helpful in examining the Trident system.

Chapter 2 concentrates on the process of technological shrinkage at VSEL Barrow and the construction of "core business," both of which facilitated the production of Trident. This largely historical chapter tries to disentangle the enormously complex developments within the huge Vickers empire.

The changing ownership and patterns of production within Vickers—one of the UK's largest engineering companies—have had a strong influence on the direction and the content of its technologies. The case of one such technology, the Constant Speed Generator Drive, is traced in chapter 3. The aim here is to document how one potentially profitable technology was marginalized in the process of constructing core business.

Chapter 4 is an account of how Vickers was taken out of British Shipbuilders and privatized in the biggest management buyout yet seen in the UK. This privatization is set in the context of the impending Trident contracts and the need to construct both a purpose-built site and a purpose-built workforce to ensure production.

The reconstructed story of the Barrow Alternative Employment Committee follows in chapters 5 and 6. These chapters explore the efforts by BAEC to generate and promote alternative technologies that could be built by VSEL without compromising its skills base, focusing on the factors that facilitated and hindered the committee's endeavors.

The heavy waves of redundancies predicted by the BAEC campaign are discussed in chapter 7, where it appears that pain and dislocation are implicit in building such a large-scale weapons system as production falls away. The theoretical issues this raises for actor-network studies are examined here and in the conclusion, where I suggest ways in which science and technology studies might address processes such as production, redundancy, and technological shrinkage.

I
Simplifying Production

1
Large-Scale Technologies: Ambivalence and Coercion

If the weapons succession process goes on unchecked, it is not because of an internal technological imperative, but because those involved are too often unrestrained by those (the rest of us) who are not.
—*Spinardi 1994, p. 194*

How do technologies succeed? In a powerful metaphor borrowed from biology, a popular model maintains that successful technological development is a "natural" process rather like diffusion. Recent work (MacKenzie 1990; Latour 1987, 1988, 1996), however, has tended to discredit this approach. Bruno Latour (1987, pp. 122–144) and other "actor-network" theorists argue that the process of creating and adopting technologies is complex, interactive, and political. To be successful, they argue, technologies must not only get built but must then get built into society. Their work aims to uncover the facts, the machines, the people, and the bureaucracies that must be aligned, molded, and disciplined to constitute a technological development. All these elements make up an "actor world," an overall environment that provides the conditions for a technology to succeed.[1]

Once a technology has been adopted and stabilized, however, the bulk of this complex, heterogeneous world often sinks out of sight, like an iceberg. In the process of creating any technology, a wide diversity of animate and inanimate entities must be enrolled (that is, persuaded to adopt identities) so that their primary function becomes that of promoting and facilitating the technology. To describe how objects, artifacts, and technical practices come to be aligned, John Law (1987, p. 111) introduced the useful term "heterogeneous engineering."

That the "right" technologies are the ones that are adopted by society is a market-oriented idea that ignores two of the central themes of this book: hidden histories and tacit knowledge. But the model of right technologies still has a powerful purchase in society. In the case of defense technology, the model is reinforced by the idea that modern weapons are inventions so complex as to be understandable by only a very few "experts." I shall argue here that we can use workers' knowledge to help construct a bridge between what sociologists have revealed about technology and how the public perceives technologies. First, however, I will trace some of the ideas about technology that have been pivotal in STS and show how these can be linked with ideas from the literature on labor history.

This study describes how the production of a large technical system was stabilized. The production process has received much attention in industrial sociology and political science.[2] Here I shall apply the insights into content offered by science and technology studies (STS) to examine the relationship between content and production. Early work in the sociology of scientific knowledge (SSK) opened up not only the practice but also the content of science, and subsequently technology, to sociological investigation. Underlying these incisive investigations was a bedrock of work in the sociology of science in which the classic norms of scientific practice outlined by Robert Merton—universalism, communality, organized skepticism, and disinterestedness—were challenged empirically and conceptually. Indeed, Thomas Kuhn convincingly challenged the image of science as a cumulative activity engaged in by scientists gradually revealing the truth about nature. Kuhn's work uncovered the power of paradigms in science, a power reflected in the tendency to blot out contingency and represent scientific development as linear.[3] This tendency helps to create scientific (and technological) icebergs, with contingency and uncertainty hidden below the surface. Today, even as the belief that science is an activity that gradually reveals Nature is questioned, it is also still very much a part of everyday discourse in the media and in the public domain. It sometimes seems that both discourses are needed to reflect the increasing ambivalence many people feel toward science and technology.

The pioneers of SSK wanted to focus on science and technology itself, and on content alongside context. Scientific knowledge came to be seen as a product of social negotiation: science was a cultural product.[4] A range of studies have shown that cultural differences give rise to differences in science—a point surely demonstrated by the poor "expert" understanding of the environmental consequences of radioactive fallout from Chernobyl in the UK's uplands, which local farmers, with their situated knowledge, were able to predict more appropriately (Wynne

1992). There is then a need for a citizen science, a dynamic link among local and global science, citizens, and scientific groups (Irwin 1995).

A range of related studies have viewed science and technology as socially constructed[5]—as outcomes of actor networks (Callon, Law, and Rip 1986; Law and Callon 1992), as expressed by paradigmatic technological systems (Hughes 1987), and as embedded within sociotechnical networks (Enserink 1993; Elzen, Enserink, and Smit 1996). This sometimes confusing array of approaches offers ways of considering the interaction of people with machines and the interaction of machines with machines.[6] Actor-network studies stand out in their assertion that nonhumans can play symmetrical parts in technological dramas. As I hope to show, giving voice to nonhuman participants can offer a way of understanding and addressing the continuing problem of technological determinism (the enduring power of the diffusion model), and it is particularly important in a study of defense technology. Because actor-network studies sometimes lack the detail and richness found within the field of history of technology; however, I use a historical and documentary approach alongside that of actor networks in this study.[7]

En route to understanding the merits of these various approaches, it is worth considering some of the history of ideas about what "technology" means. In his discussion of technology and politics, Langdon Winner (1986, p. 40) points out that *techne* (from the Greek, meaning art, craft, or skill) was used by Plato to mean statecraft or policies. This definition is particularly useful in underpinning a study of weapons systems. For Jacques Ellul (1964, p. 125), *la technique* signified the use of rational methods to achieve outcomes in any field of human activity; in his view, this severe rationality carried the constant threat of totalitarianism. From these philosophies of technology, others have developed the concept of "technopolitics"—the idea that doing technology and doing politics are one activity.

Another influential tradition explores technology as the embodiment of knowledge, both explicit and tacit. The unspoken nature of the knowledge that is bound up in technology was described by Michael Polanyi (1967, p. 18): "It is not by looking at things but by dwelling in them that we understand their (joint) meaning." More recently, Harry Collins (1985), in his work on replication, explored the pivotal role of tacit knowledge in science, while Donald MacKenzie and Graham Spinardi (1995), in their examination of nuclear weapons proliferation and the possibility of "uninvention," carried this analytical work into technology. In the case of this book, the search for tacit, hidden knowledge involved seeking out histories of Trident production that emanated from a rich variety of sources. Reconstructing some of these histories offers the chance to uncover knowledge that predated and may also survive the technical system it helped to create.

There are two further aspects of uncovering hidden knowledge that are important for this study. First, Trident (the submarine and the weapons system) is a secret technology. An in-depth study of its physical construction would therefore have been unrealistic because this would have touched on areas deemed to involve national security.[8] Researchers in this field have commented that there is a more pervasive climate of secrecy and self-censorship in the United Kingdom than there appears to be in the United States (Donald MacKenzie, personal communication). Much of this study, therefore, focuses on the generic nature of the technical know-how that came to be embodied in Trident, but which otherwise could have been channeled into other products. Second, essential to an understanding of how this large technical system came to be produced is an account of technical contingencies, which include potential technical "distractions" and political uncertainties surrounding production, all of which somehow had to be managed and overcome. Many of these contingencies were explored in the unofficial accounts given by workers.

Roads Not Taken

The concept of technological roads not taken is fundamental to my approach.[9] David Noble's social histories of technology underline the importance of looking at both what is produced and what is not produced. Much as Latour rejects diffusion, Noble rejects what he calls the dominant "Darwinian" view of technological development, where only the "best" or "fittest" technologies get produced or survive. Such a view of technology relies, he argues, on the belief that it has passed through three successive filters: the "objective technical filter," the hard-nosed filter of "economic viability," and the "self correcting mechanism of the market" (Noble 1984, p. 145). But, Noble continues, this rationale is facile and cannot account for technology's "people, power, institutions, competing values or different dreams." Existing technologies have rarely been put through "natural selection," whose series of tests are really political and cultural. If an alternative technology challenges the "established way of doing things," if it does not fit in with the predominant technological scheme, it is then subjected to the apparently rigorous filters described above. Since these filters are political and cultural rather than technical or economic, such alternative or less conventional (but possibly, in the commercial sense, more dynamic) technologies will fail.

Mary Kaldor's exposition of the degenerative effect of military contracts on what gets counted as technological innovation views Trident as embodying the extreme, the pinnacle, of "baroque technology." Her analysis of the "mummification" of UK industry, which had previously centered around shipbuilding and steelmaking, underlines how certain industries, notably (naval) shipbuilding, were allowed to grow within the confines of outmoded industrial configurations, creating an artificial sense of prosperity. "Baroque technology" describes a process whereby essentially conservative technologies, such as the naval submarine, are subjected to ever more elaborate and sophisticated "improvements," either in design or in components or subsystems. What results is a contradictory, flip-flop approach to technology. Trident systems can be presented as at the cutting edge of technical development, yet, as I hope to show, they can hold back and stultify technical dynamism. Kaldor (1981, p. 24) notes that "the outcome of this contradictory process, in which technology is simultaneously promoted and restrained, is gross, elaborate and very expensive hardware." From here it is a very short step to those studies of science and technology in which no distinction is drawn between the technical and the social, between technology and politics. (Thus, the filters described above are technical and economic at the same time as they are political and cultural.) But (and this is both a philosophical and methodological point) it sometimes becomes necessary to distinguish the two when *writing* about technology—to describe and explore how the terms "technical" and "social" get used, in order ultimately to reunite their meaning.

The STS Toolbox and the Construction of Trident

Because the starting point for this study was the ambivalent nature of the relationship between shipyard workers and the Trident submarine construction program, theories that went beyond viewing technologies as configurations of successful, stable artifacts were of immediate interest. A particularly helpful tool was the social construction of technology (SCOT) approach of reopening technical controversies through detailed case studies to reveal how technologies went through stabilization and closure.[10] This approach usually reveals that apparently stable technologies started with many possible futures and have been shaped by particular social interests and relevant social groups and interpretations.

In view of the program's huge scale and its local, national, and global aspects, a sociological study of the nature and purpose of Trident submarines calls for theoretical models that allow for multi-stranded accounts and recognize an heterogeneous array of sources. My account needed to allow for a strong sense of organizational and bureaucratic as well as technical pressures on Trident production. I found that Thomas Hughes's notion of "reverse salients," describing what happens when a technical system hits problems or obstructions, lends itself to the study of large-scale, complex technologies.[11] This is particularly true for large sociotechnical systems such as Trident because Hughes's reverse salients can be both technical or organizational in character. Often, roads not taken by a company that was moving toward national monopoly of submarine building were not taken for organizational reasons rather than for "technical" reasons.

Because some Trident workers were ambivalent toward the weapons system itself, the concept of "enrollment" was crucial to this history. Thus, I have drawn deeply on the actor-network approach, from which the concept of enrollment principally developed. The term "enrollment" as used here refers to "the definition and distribution of roles by an actor world. It should be noted that roles are not fixed and pre-established, and neither are they successfully imposed upon others" (Callon, Law, and Rip 1986, pp. xvi–xvii).

I attempted to explore enrollment in a way that would avoid the "heroic" bias of many STS accounts and their fixation on the originators of networks.[12] While the goal of many researchers in STS has been to write unheroic accounts, "big" actors have often stayed at center stage in their accounts. With the story of Aramis (Latour 1996), the technology itself moves to center stage. Another way to write an account that was not focused on management was to try to understand the simultaneous disempowerment and disenfranchisement of workers, and the violence done to knowledge and machines, that can be brought about by enrollment into the production of large technical systems.[13] Law has urged that the traditional sociological concern with the distribution of results should become more of a feature of science and technology studies—we should, he says (1986), study "heroic failures."

The relationship between technology and unemployment is a major feature of the "pain and dislocation" that can result from technological change, but this connection has often been left unexplored in STS.[14] Labor and the production process are entirely missing from Spinardi's 1994 study of the Polaris-to-Trident story and from MacKenzie's 1990 exploration of missile accuracy, yet these studies are otherwise richly contextual and are fine examples of the analysis of the relationship between society and technology.[15] Sociologists and historians of technology in the Marxist tradition have written about the redundancy and de-skilling resulting from the introduction of new technologies and the increases in managerial control brought about by the introduction of new technical systems,[16] but STS has been largely silent on those topics.[17] Andrew Webster has argued that scientists and white-collar technologists should be regarded as producers in order to correct the tendency of the sociology of science to ignore production and the relationship among science, technology, and the labor process. Webster (1991, pp. 95, 101) also points out that power, choice of technology in production, and wider work relations have had little impact on the sociology of science.

Large-scale weapons technologies require large, multiply skilled workforces for their construction. Once they are built, though, thousands of workers will inevitably be jettisoned from the network. These highly specialized workforces must therefore live with the constant threat of cancellation or "downsizing." Useful and incisive as they are, STS and actor-network studies tend to describe the buildup but not the build-down of networks. I have used the term "disenrollment" to describe the process of ridding networks of people (as well as knowledge and machines); in this I am making use of the tools laid out by the actor-network approach to the study of technology, such as remaining "undecided about what technoscience is made of" and remembering that "to understand what facts and machines are is to understand who the people are" (Latour 1987, pp. 258–259). Susan Leigh Star (1986) has pointed out that "every enrollment entails both a failure to enroll and a destruction of the world of the non-enrolled." In the case of Trident, however, we witness a strategic marginalization rather than Star's "accidental" one—a proactive marginalization as opposed to a more passive failure to incorporate.

Revealing Production

Sociological studies of technology tend to underplay the production process, as if it were something given and unproblematic that needed no explanation. Inputs are studied and outputs are observed, but the process by which input becomes output remains occluded.[18] But surely actor worlds must also be built and sustained in order to bring a large-scale technology into production, and these are strongly linked to the actor world being prepared to receive the technology and allow it to operate. Stable, reliable production is particularly important when the technology is a complex weapons system in which an element of exclusivity or secrecy is involved and perceptions about (technical) workability are volatile. The production process itself must be stabilized and secured, and uncertainties must be removed. That such technologies will or can be produced is almost as important to "technical" success as their commissioning and use.

Production problems can destroy credibility.[19] Both industrial relations and detailed quality control were crucial components of Cold War credibility. I will discuss two major UK shipyard strikes, much as Tyler (1986) describes labor problems, shutdowns, and brinkmanship between General Dynamics and the US Navy.[20]

Securing the production process is also crucial for large technical systems in which workers, components, and other variables could start to dissociate themselves unless kept in line by a process of continually having their interests shaped or (re)connected with the system; the actor-network approach calls this process "interessement" (Callon, Law, and Rip 1986). The particular circumstances of the UK's Trident shipyard privatization, related in chapter 4, will illuminate this point.

The actor-network approach also emphasizes the local dynamics of technical development and the micropolitics of enrollment, both of which are pertinent to this study. Michel Callon (1987, pp. 95–96), describing the attempt by Renault to promote an electric vehicle, the VEL, refers to "a black box that contains a network of black boxes that depend on one another both for their proper functioning as individuals and for the proper functioning of the whole." Here Callon is talking about the relationship between "individuals" (components of the VEL such as hydrogen or fuel cells) and the "whole" (the VEL itself). There is, however, a sense in his description in which each entity, such as the fuel cells, has its own dynamic, or actor world. We could develop this further: the Trident "production actor world" could be seen as an entity comprising diverse elements—workers and workshops, promotional literature and gear teeth, welding torches and managers, computer-aided design systems, and trade unions. To help us understand how and why production succeeded, we could view production as an actor world in its own right before relating it back and forward to the huge, interdependent, and international defense system in which it belongs.

Technopolitics

Trident is a political technology in Winner's sense. If it takes a society that embodies a certain kind of (totalitarian) politics to make a system such as Trident a success or to allow a technology such as nuclear power to be regarded as "safe," then the weapons system and the nuclear power plant come to exemplify those politics. It is not that the politics precede the production, but that each creates the other. Winner (1986, p. 17) asks "As we make things work, what kind of world are we making?" But if we are committed to the concept that technology is social to the core, can some technologies really be political while others are not? Trident is perhaps an extreme case: If you were to ask people Winner's question "Do artifacts have politics?" they might reply "No" when it comes to bicycles but "Yes" when it comes to a submarine-launched ballistic missile. Trident's politics are overt. But does this mean that Trident is a special case and that other technologies can be seen as apolitical? Perhaps this notion that some forms of technology are politically neutral accounts for the persistence of the neutrality myth—the idea that technology is essentially value free but occasionally gets corrupted or used as a tool in an "evil" game from which other "evil technologies" then spring.[21]

However, some technologies really do appear simply to be tools, simply to be neutral, while others definitely do not. Asked about technology in the abstract, rather like science in the abstract, people may more easily describe it as something neutral, pure, and untainted than if the question is focused on particular (overtly destructive) artifacts or technical capabilities such as submarine-launched nuclear missiles. The answers to these questions will vary enormously according to the social world inhabited by the respondent. Further, some people working on SLBMs may maintain the neutrality myth (possibly for their own psychic survival if they have deep misgivings about the technology) by separating the whole technical system into neutral-looking and more morally manageable components or subsystems. Others may simply not see it as a part of their role to question or engage with the content of production. (The political scientist Albert Hirschman described these kinds of strategies within an organization as Loyalty.) Or, in the case of the group of shipyard workers whose story is told in later chapters, insiders may decide to remain in the network but work to change it. They see the network as partly theirs, and so theirs to change. They may therefore seek to change or influence the content of production (a process described by Hirschman as Voice). Others may quit or Exit (Hirschman 1970). In the climate of an economic recession, and certainly in the case of Trident's now-declining network, the most populated of these categories among submarine workers has tended to be Loyalty. This also appeared to have been the experience with the US program during a period of mass layoffs.[22] But more explanation of how this "loyalty" came about is needed if the success of Trident production is to be understood. The second-most-populated category was Voice, whose articulations are explored throughout this book.

The politics of certain technologies remain overt. The Trident system doesn't appear much like a tool that can be corrupted to evil use. Indeed, its use at all would be a global disaster. Nevertheless, it is still surrounded by powerful myths of inevitability that tend to support a kind of intellectual determinism: "You can't uninvent the bomb." But large-scale weapons systems require huge socio-technical networks and still employ large numbers of skilled people who must remain committed to the project. Does it follow that all those employed in the defense industry somehow suspend political judgment and ignore the possible outcomes or consequences of building these technologies? If this is not the case, how are they enrolled into such networks?

The somewhat murky idea of political technologies appears to sit uneasily beside the antiseptic actor-network approach to technology, which accounts for technical success as an outcome of stronger or weaker associations, stronger or weaker networks or constructions. However, to build an actor world that enables technical development, and to enroll people into that system by shaping them and their interests to converge within the project, is to do politics. Winner argues that some technologies are inflexible and coercive in some circumstances but not in others. Spinardi points out that success in technology depends on control, and that the larger and more complex the technology the more critical is the need for that control and the greater are the penalties for losing control. All large-scale technological developments threaten democracy, argues Spinardi (1997), because their success depends on the control of much more than narrowly defined "technical" issues. But if a dark technology such as Trident can also be seen as a collection of more benign technologies and processes, the interesting question is how these technologies and processes became configured to produce such an inflexible, coercive, and overtly political technological system.

Technological Determinism and Nonhuman Actors

The idea that technology might be autonomous, somehow containing an internal dynamic and unaffected by exterior forces, partly explains many people's sense of disempowerment and passivity in the face of technical change. In their discussion of technological determinism, MacKenzie and Wajcman (1985, p. 4) draw interesting parallels with nineteenth-century notions of climatic determinism—powerful forces of nature with which argument is futile. In spite of the large critical literature on technological determinism,[23] this idea retains potency in the defense sector, where the discourse of technological trajectory in weapons building is supported by other determinisms (political and economic rationales of inevitability) that underpin the "upgrading" of weapons systems. Spinardi (1997) gives a useful example of the belief that technology itself has driven the arms race in his trenchant study of the social, political, and technical elements involved in UK nuclear weapons development. He quotes from paragraph 67 of a 1981 Report of the Secretary General of the United Nations titled Comprehensive Study on Nuclear Weapons: "It is widely believed . . . that new weapon systems emerge not because of any military or security considerations but because technology by its own impetus often takes the lead over policy, creating weapons for which needs have to be invented and deployment theories have to be readjusted."

Many examples of the view that weapons have trajectories will be found in the following chapters. Throughout the 1990s there was also a strong sense of "economic determinism" among workers in the defense industries, which were seen as the last outpost in the UK of manufacturing, technical innovation, and industrial employment. The major defense-industry trade unions launched a series of initiatives in the early 1990s to combat this sense of economic determinism, but their arguments were largely ignored.[24]

One important dimension to deterministic thinking in the defense sector relates to the scale of the technical systems involved, which often influences the level and type of analysis undertaken. Thomas Misa has argued that machines are allowed to "make history" only in studies conducted at the macro level. Thus, philosophers of technology who adopt a macro-level perspective often lean toward determinism, whereas labor historians, who often adopt micro-level analyses, typically deny the thesis. Interestingly, Misa (1994) found that historians of business, the city, the physical sciences, and technology tend to take intermediate positions. Histories of technology that trace in detail why and how artifacts look and behave as they do are of great significance in undermining the status of the machine. Noble (1984) has shown how technologies in the United States became systematically militarized, which in turn shaped both technical development and power relations between workers and management. In his study of how American business bureaucratized science and technology, he showed how important it was to this process to eliminate any sense of contingency and uncertainty in the innovation process, in effect creating technological icebergs. Part of this process involved use of the patent system to reduce or expunge the role of the individual inventor in favor of corporate innovation (Noble 1977).

It seemed important in writing this book to investigate the popular persistence of determinism as an explanatory theory of technical change. Social-constructivist theories about technology, while powerful, have not succeeded in permeating popular or policy thinking in Western society. Policy makers can and do still claim that "there is no alternative." It seems difficult for sociologists to grasp the materiality, the physical presence, of artifacts—to achieve, as Law and Bijker (1992) put it, a dialectic in which people and circumstances are balanced. Though the myth of technological trajectory can be teased apart, the fact is that, no matter how strong our case that technologies and technical systems are social constructions, once constructed they do, by their existence, demand accommodation. Even Winner (1986, p. 6) acknowledges that "technologies provide structure for human activity" and that "they are not merely aids to human activity but are also "powerful forces acting to reshape that activity and its meaning."[25]

But we are not technological determinists just because we recognize that technology can have effects. The question is one of degrees of obduracy. Part of this study details a process of technological shrinkage that made the construction of Trident possible. Once that shrinkage was accomplished, the possibilities for technological diversification (to retain employment) became virtually nonexistent. We could argue therefore that Trident had irreversible "effects," such as destroying future chances of employment. Bijker and Law (1992, p. 297) put it this way: "Much of the process of barrier building has to do precisely with distinguishing between who will be inside and who will be outside . . . the allocation of rights and duties. Often these have to do with the rights to speak, or the duty to keep silent—a process which involves disenfranchising those who find themselves on the wrong side of the barrier. To the extent that those outside depend on or have an interest in the product, the product and its producers become an obligatory point of passage."

What emerged from my localized study of Trident production was the recovery of the right to speak and the importance of avoiding passivity when thinking about, using, and building technology. To say that Trident caused unemployment would perhaps be simplistic determinism, but to say that Trident is the physical embodiment of the process of technological shrinkage and job loss, and that building Trident was doing "politics by other means," could well be useful.

Because of the importance it places on the histories of nonhuman components of technology, the actor-network approach can help overcome this philosophical problem of the physical presence of artifacts. Perhaps the claim that technology has effects is another way of saying that nonhuman actors can have an influence on the course of technological developments. By allowing nonhuman entities to have "life," to be a real part of the story, we can come closer to understanding the influence of artifacts. Powerful narratives that give a sense of the drama of technology can be told in this way.[26] The nonhuman actors in this book include drawings, boiler shops, and bed plates, whole machines and parts of machines, knowledge of various sorts, skills, and policies. They are sometimes phantoms—intermediaries from the past who return to haunt the present. The Trident production process, and the maintenance of the system's "workability," required the manipulation of tools and components as well as the enrollment of workers and macropolitical and micropolitical stabilization. Artifacts, systems, people, and perceptions were all crucial to the program's success. Production of some of these components afforded high prestige to designers, managers, and sometimes workers, where these involved new, untried technologies. Production of Trident involved unique applications of skill because of the nature of the content of some of its component technologies. My goal in this narrative will be to avoid determinism by studying the relationship between skill and technical content, which should make it possible to "unpack the black box" of production.

Labor and Disarmament Studies—Content and Context

The literature of human-machine interaction and human-centered design includes studies of earlier attempts, often by workers, to open up production and technical content to social and political scrutiny. This has most often occurred in defense industries where workers have campaigned for conversion to "peace products" or "socially useful technologies." Studies of technology drawn directly from the shop floor, often reflecting the experience of de-skilling through the imposition of technologies that removed worker control, were early examples of questioning the human-machine interaction, of reuniting the content and context of technologies and recognizing that changing or influencing the content of production involves a radical shift in power relations. Mike Cooley's *Architect or Bee?* (1980) and the influential Lucas Plan (on which see Wainwright and Elliot 1982) are perhaps the best-known examples of this movement against determinism by workers. There have been other workers' campaigns in the UK, not only at Barrow in Furness but throughout the old Vickers empire, starting with attempts to establish a workers' combine in the 1970s.[27] There have been initiatives at Vickers on Tyneside and more recently at Ferranti (later GEC) in Edinburgh. The latter initiative helped spawn the national Arms Conversion Project, a Glasgow-based alliance of workers, trade unions, and local authorities pressing for employment-saving alternatives to defense dependence.

The Lucas shop stewards' alternative production plan was an attempt to unite the content of technology with its context, the technical with the social. Rather than being exclusively producers of technology, the workers also wanted to play a role in its creation and design. Workers' alternative plans occupy a long (albeit minority and unofficial) tradition within the trade-union movement. Underpinning these plans has been a process of education about technology and power stemming from local and regional trade-union activity. A good example was the site-by-site negotiation over the introduction of new technologies in the newspaper industry. After the workers' initial defeat when direct input technology was imposed at News International's Wapping newspaper complex in the 1980s, ground was slowly recovered by active local engagement and deconstruction of these technologies to reveal health, safety, and employment implications. The trade unions were then gradually able to influence content as well as context in the workplace. Though the extent of this influence should not be overstated, I would argue that the alternative-plan movement has been insufficiently recognized and understood within science and technology studies.

The Role of Absence

Both identity and power are important features of any actor-network account of technology.[28] The present study involves exploring what was not produced as well as what came into full production and development. The search for what was not produced provided a way of discovering power relations that could not be seen by observing the recent technical outcome or output of the shipyard. Steven Lukes's analysis of the configurations of power remains pertinent. Lukes (1974) calls studies of power that concentrate on behavior and observable conflict one-dimensional and limited. He argues that we should examine "non-decision-making" alongside decisions, "potential issues" alongside issues, and "latent conflict" alongside conflict. He then suggests that we go further and ask how agendas are shaped, uncovering a landscape beyond non-decisions and potential issues where decisions and issues are not perceived and do not arise. Looked at this way, even the absence of conflict tells us something about the actor world that has engendered that absence. Contemplating the role of absence, particularly in the context of redundancy, Mike Michael and I have developed the concept of absent intermediaries (Mort and Michael 1998). This concept will be explored in later chapters.

Conclusion

In the early 1980s, it seemed that the Cold War would go on for the foreseeable future. That was the environment in which the Trident system builders worked. The East-West standoff seemed to have taken on an autonomous life, rather like the predictions of ever increasing demand for electricity described by Hughes (1987)—predictions that then had to be satisfied. A trajectory of ever bigger, more destructive, and more accurate nuclear weapons and delivery platforms was devised to accommodate the standoff and the consequent defense strategies. The builders of such large-scale technological systems had to be heterogeneous engineers, in the sense that they needed to call upon a wide variety of skills to manipulate a broad range of technical, scientific, economic, political, and social elements to ensure that the technology they promoted would be seen to "work." During the Cold War, the perception that a weapons system "worked" was of the utmost importance. A threat to production could also be a threat to workability: "Establishing the credentials of a nuclear deterrent system, therefore, begins with its procurement; maintaining these credentials continues throughout the system's life cycle." (Dillon 1983, p. 135)

Those who sought to question or deconstruct Cold War weapons systems also needed to be heterogeneous engineers. By following the actors in the Barrow shipyard, I was able to rediscover the heterogeneous engineering by which a group of employees hoped to steer the shipyard away from Trident production and into alternative marine-based technologies.

At the same time, I could build a picture of the heterogeneous engineering by which the Trident builders ensured that the alternative, distracting technologies promoted by those workers could not be developed.

The actor-network approach reveals the heterogeneity within the apparent coherence of successful technologies.[29] But missing from the list of actors who can create or kill off new technological systems are a series of others who have had to be excluded for the new system to cohere. These others are unemployed workers, silenced engineers, dismissed critics, and also other technologies and roads not taken. This study focuses on the missing and the silenced.

2
From Diversity to "Core Business"

We used to do cement, ink machinery, soap machinery, condensers, boilers, it was all commercial. . . . I was in the boiler shop and the place was packed with all types of commercial equipment. Now it's like a ballroom, it's empty, they've taken the machines out and they're just knocking the plant down now. They just knocked the training school down; we had two. . . . You see, you can't run a company if you're not training people.

—*from interview with a retired chartered engineer who had been chief designer, chief contracting engineer, and finally development engineer in Vickers's cement division, December 3, 1993*

The story of the gradual transformation of the Vickers Barrow engineering plant and shipyard from a heterogeneous civil and military facility into a single-purpose nuclear submarine construction site is important to our understanding of the fixing of the Trident production process.

This transformation was influenced by changes in the overall ownership and structure of the company that resulted from policy shifts by alternating Labour and Conservative governments. Parts of Vickers's industrial empire went through nationalization in 1951, denationalization in 1954, and renationalization in 1977; the shipbuilding section was again privatized in a management buyout in 1986.[1] These politically driven changes had a major effect on technological content and on the possibilities for industrial synergy.

In addition, the company's decision to pursue single-purpose defense production, culminating in the production of Trident, secured employment in the short term but can now be seen to have brought about the destruction of engineering jobs in the medium and the long term. As the Trident production program began to wind down, the decision to specialize became associated with loss of employment. It was argued that if the company had retained its technical heterogeneity it would have kept open some commercial markets that might have become a springboard for new enterprise (BAEC 1986). As part of its campaign for jobs in the early 1980s, the Barrow Alternative Employment Committee had to assemble arguments against the attractive short-term economic security offered by the Trident program. At the same time it found itself arguing against the steady process of "technological shrinkage" within the company.

It has been suggested, notably by Steven Schofield, that the nature and scale of submarine production at Barrow was such that by the mid 1980s diversification into (or back to) other products was not feasible. Schofield advocated a community-based rather than an industry-based approach to defense diversification, involving macro-level initiatives such as promoting inward investment and more regionalist economic policies by the central government. This would provide a chance for reducing defense industrial dependence.[2] However, the accounts given by workers and managers in my study suggest that the story is not one of a company having a homogeneous identity and needing to break into other markets but one of a company steadily *divesting* itself of successful, often profitable products. This divestment was part of constructing the concept of "core business" in a contracting defense market. "Diversification" in the Vickers Barrow context is a slightly misleading term.[3] Though there probably came a point for the company by the end of the 1980s at which recovering lost markets was no longer possible (the commercial "edge" having been lost and social/technical contacts dissolved), nevertheless a vibrant, enterprising, and diverse technological base (and workforce) had only recently been sacrificed to the single-purpose industry, the "core business."

All the other (mostly commercial) technologies discussed in this chapter disappeared from production, marginalized to strengthen the emerging core business. Managers, designers, and workers gave accounts of this process of technological shrinkage, revealing a multiplicity of ways in which these hybrid, largely commercial networks were weakened. What emerges from these accounts is a sense of the gradual but steady decline in the perceived status of civil commercial business. The robustly stated company policy of constructing core business through the 1970s and the 1980s was, however, initially softened during the worst of the defense cuts and the redundancies of the early 1990s. This was briefly reflected in statements about a policy of "limited diversification"[4] and a hope that 25 percent of the company's turnover would come from commercial products by 1995.[5]

Whether to concentrate on "core" products or to "diversify" was not a new management problem in the company's history. Originally a steel company, Vickers moved into defense work in the 1880s as the UK's defense spending grew and as protectionism limited the export of steel to overseas markets. Surges in defense production for the world wars set a pattern that increasingly pushed civil work into the role of peacetime filler. One history of Vickers notes the effect this had on production: "Civil products at Barrow and Elswick in particular were abruptly stopped and thrown off the production lines; negotiations for new engineering products, which had been delicately nursed for months, or even for years, were canceled overnight, and some disgruntled customers never came back." (Scott 1962, p. 357)

Although Vickers to some extent became used to "boom and bust" in the swings between war and peace production, overdependence on arms production was nevertheless often seen as risky. Management consultants brought in to appraise the company in 1963 reported: "Our dependence on the armament business is as high or higher that at any time in the last 4 or 5 years (32.1 percent of total group productive wages). This does not mean that we want to reduce our armament business, but only its significance in relation to the total."[6]

Recently classed as the world's most defense-dependent company, VSEL devoted more than 95 percent of its production to defense in 1994 (Finnegan 1994).

Large-scale arms production during the world wars became irrelevant when hostilities ceased. Both wars were followed by successful diversification drives by Vickers, but not before the company went through rounds of mass layoffs. In 1917, for example, the Barrow workforce stood at 31,000. By 1923 it was down to 3769, with 44 percent of local men unemployed (Trescatheric 1985). Searching for new lines at the height of war production booms was not considered necessary or possible. It could be argued that the scale of the Trident program brought its own specialized form of employment boom, mimicking wartime production, and that it was anticipated that the Trident contracts would prevent diversification and technological heterogeneity.

The alternative-employment campaign of the mid 1980s was an attempt to break this boom and bust pattern by securing alternative sources of production during the period of financial strength offered in the short term by Trident. As Trident production wound down, the "baroque" nature of the technology—its political, social and technical characteristics—precluded any other form of production. Although some of the Trident submarine's *component* technologies might possibly have commercial synergies or potential applications, these cannot now be exploited because the social or market-oriented avenues to commercial production were closed off in the process of strengthening the dominant network.

Examples of technological shrinkage include the case of the Constant Speed Generator Drive (discussed in the next chapter), the closing of Vickers's cement division, and the whole downgrading and decline of general engineering at the Barrow plant. This chapter will describe the scale and range of engineering work once undertaken by the company through the accounts of workers. These accounts form an alternative history of the company and of the relationship between defense-oriented production and employment.

Monopoly or Diversity?

VSEL Barrow came to be perceived in the 1990s as a one-product shipyard: "the nuclear submarine maker VSEL."[7] But this role was a comparatively recent construct. There had been frequent periods in the history of both the Vickers empire and specifically the Barrow plant when its general engineering activity was more prominent and more profitable than its shipbuilding. In an interview (January 13, 1995), Hugh Kelly, who worked at Vickers for 46 years, starting as a apprentice fitter, then commercial draftsman and finished as manager of the Commercial Department in the late 1970s, said: "There was a clash of jobs in there. There was armament, shipbuilding—that sort of thing—and then commercial . . . deadly rivals in the boardroom. . . . Commercial was always the poor relation, but we pulled our weight and made a heck of a lot of profit . . . but we couldn't get access to money for design work, research for high tech development." At its peak the Vickers empire encompassed four main activities: steelmaking, shipbuilding, engineering, and aircraft. The Barrow plant was historically a hybrid within Vickers because it had a strong base in steelmaking, engineering, and shipbuilding. The early 1960s saw the profits of the Vickers conglomerate melt away and its vast empire shrink, largely because of the nationalization of the UK's steel industry in 1964. In addition, Vickers's aircraft division was experiencing severe problems with the development of the VC10, and it was acknowledged that the drain of financing that project had led to the impoverishment of the company's general engineering activities (Evans 1978, p. 87).

Shipbuilding and engineering are closely related activities; in fact, in the 1960s Vickers's annual accounts listed them under a single heading.

In some cases the separation of engineering and shipbuilding rested on nuances (particularly at Barrow, where much of the engineering work was related to shipbuilding). But there were also definite and separate activities, where self-contained contracts had been won or where Vickers had simply acquired existing engineering companies. For internal company accounting purposes a line came to be drawn between the shipbuilding and engineering sections, and in three successive years (1964–1966) Vickers shipbuilding nationally incurred a loss (although shipbuilding at Barrow remained marginally profitable).[8] But within the Vickers industrial empire (both globally and nationally) it was the Barrow Engineering Works that maintained a "high order book" and contributed most of the profits. Barrow engineers alone made £2 million in profit in 1961 and again in 1963. The manufacturing of ships' engines, installations of ships' machinery, and armament work provided the basic workload. But it was the massive contracts with British Rail for manufacturing Sulzer diesel traction engines under license at Barrow that had the greatest impact on profits. These contracts have become legendary in Barrow. There were, for some, high wage bonuses to be made on production of the Sulzer engines.[9] Other non-defense work included the largest ship engine built to that point at Barrow: a 17,600-horsepower diesel for the *Australian Star*. Harold Evans (1978, p. 91) adds: "Cement plant and industrial pumps were other important Barrow products, but both were 'fading,' due probably to lack of development effort." This is one of the very few references Evans makes to engineering at Barrow, even though the plant's non-shipbuilding work was such a large earner. By the late 1960s Evans already perceived Barrow as primarily a shipbuilder. Defense-related engineering work at this time involved the design and development of missile launching systems, missile tubes for the Polaris program, and installations at the new Dounreay nuclear research station in Scotland.

Engineering profits again overtook shipbuilding profits in the mid 1970s. The Vickers UK engineering group made a £6.2 million trading profit in 1976, and with Vickers's Australian and Canadian engineering companies added in the total was £12.1 million in that year. Also highly successful were the office equipment and lithographic plates businesses (located in London and Leeds).

During the 1970s, the status of engineering within the Vickers empire was inextricably linked to the nationalization plans of the Labour government elected in 1974. This government was committed to nationalizing the aircraft and shipbuilding industries. How industrial content and activity were defined became of utmost importance for Vickers; indeed, the distinction between shipbuilding and engineering became crucial to "protecting" parts of the business from nationalization. Engineering was, in that sense and at that time, less politically uncertain. But because the profitable engineering businesses, based in the Barrow Engineering Works, had been taken into the Vickers Shipbuilding Group during a "power struggle" in 1968, they were therefore included in the 1977 shipbuilding nationalization and lost to the Vickers empire with its other general engineering networks. In this way the fortunes of Barrow's engineering projects became inextricably linked to shipbuilding. In addition, there is also some suggestion by Evans that, in the industrially uncertain three-year run-up to nationalization in 1977, there was less incentive for Vickers to invest in the shipbuilding side of the company as a whole because it was destined to be taken into public ownership. The Barrow shipyard and consequently its diverse engineering works suffered from this. In the context of the catastrophic national decline in the UK shipbuilding industry that followed, shipbuilding at Barrow was later "strengthened" when the plant became the prime contractor for the UK's nuclear submarine fleet. The result was an apparently secure, if dependent, internal market for nuclear submarines, but Barrow's engineering works had no such support, having to compete in the wider commercial market.[10]

In the second half of the 1960s, Vickers radically pruned its activities, selling many of the businesses it had acquired under its earlier policy of diversification, either because they were held to be unprofitable or because they no longer "fitted" together. Nationally the company pulled out of hovercraft development, tractors and earth movers, tabulators and computers, railway and rolling stock, and (later) chemical engineering. Shipbuilding and ship repairing was contracted in the Northeast, but not at Barrow. Evans (1978, p. 157) notes briefly that "the industrial pump business at Barrow was closed," but he gives no specific reason, and he does not mention the closing of the cement division at Barrow. But there were also new engineering-related acquisitions. For example, Compact Orbital Gears (Rhayader, Wales) and Slingsby Sailplanes (Kirkbymoorside, Yorkshire) were bought up by the Shipbuilding Group.

It seems most likely that Evans (who was, after all, writing a company history celebrating its survival of the 1977 nationalization) largely ignored Barrow engineering because it had been absorbed by shipbuilding, was then nationalized and therefore no longer of interest. The variation in emphasis between the recollections of former workers and the "official" histories such as Evans's[11] suggests that what was still significant engineering activity at Barrow became "invisible" both within the company (where it had been severed from the national Engineering Group) and to the "outside" world (where the Barrow yard was becoming more and more closely identified with naval shipbuilding, and particularly with nuclear submarines and the Polaris contracts). To simplify: Barrow engineers disappeared into shipbuilding, and shipbuilding then dissolved into naval submarine building.

General Engineering

In the 1977 nationalization, Barrow Engineering Works was taken, along with Vickers Shipbuilding Works (Barrow), into British Shipbuilders. Under nationalization, the Barrow plant was (to widespread confusion) allowed to keep its old name, Vickers. Even though the plant was now a part of a nationalized shipbuilding company, shipbuilding and engineering locally were to retain their separate structures and to some extent their own identities. This demarcation reflected a long-standing rivalry between shipbuilding and engineering at Barrow, where engineering had often enjoyed higher status. A young apprentice who later finished up as a senior VSEL manager described this sense of status in an interview on June 30, 1994: "In those days you had choices. . . . I chose to go into the engineering works because . . . I saw the shipyard as pretty limited in scope, whereas in the engineering works—there was a lot of rivalry and still is, those of us that are old enough—you had . . . the role of the engineering works was it was a company in its own right. . . . Apart from doing work for here and shipyards around the country, they also had various commercial products like cement machinery, soap machinery, diesel engines, a whole range of engines, there was a hell of a variety in the engineering works, it was more interesting." Technical brochures were published describing the activities of the two wings of the Barrow company under the new ownership. One of these, a glossy pamphlet titled Engineering for Quality, describes the three divisions of the Barrow Engineering Works: Manufacturing, Armament, and Mechanical Engineering. Engineering for Quality gives an account of the relative strengths of engineering at Barrow and of the position of engineering within the whole Barrow enterprise. The Engineering Works was producing guns, machinery for nuclear submarines, and missile launchers, but it had "retained a 30 percent involvement in commercial and specialized engineering products" (Vickers Shipbuilding

Group Ltd. 1977). This "30 percent involvement" was merely the rump that remained after the abandonment by the company in the early 1970s of some of its biggest commercial enterprises—in particular, its Barrow-based operation that manufactured cement-making machinery. Vickers had taken a world lead in that field in the course of its post-World War I diversification drive. Other important operations had been the manufacture of circulating water pumps for irrigation and sewage and the design and manufacture of mining engineering equipment for the world market. The manufacture of equipment for power stations was also big business. Whole units of condensers and feed heaters would be made and installed in new 30–120-megawatt power stations all over the world. Each power station unit was purpose designed in Barrow.

Even after the official closing of the cement division in 1970, commercial engineering was still the largest component of the Barrow Engineering Works' annual turnover of £30 million. Of this, general, commercial engineering contracts accounted for £10 million. Next came submarine logistics at £7.5 million, with the remainder divided among nuclear and conventional marine engineering, armaments, and spares. This 1977 booklet shows how substantial a role was still being played by commercial contracts at Barrow.

It was possible to have worked in the Barrow plant for half a century and never have had any contact with shipbuilding or defense work. In an interview conducted on December 13, 1993, a former chief designer in VSEL's cement division said: "I worked 50 years in Vickers and I never went on a boat. . . . The only time I went on a boat was in 1942 when they were building the *Indomitable*, it was just a shell and they used it as an air raid shelter, and there was an air raid."

The Closing of the Cement Division

It is difficult to overestimate the scale of Vickers's cement business. The Barrow Cement Machinery Division, shut down in 1970, was at the center of a worldwide network of companies and manufacturing plants. The cement division in some ways mirrored the Vickers empire, with satellite companies set up to operate semi-autonomously in Australia, Argentina, Canada, India, and South Africa under agreements designed to avoid intranational restrictions on imports and heavy freight charges.

Contracts ranging from those for single pieces of machinery to "turnkey" contracts for complete cement plants were undertaken. Barrow manufactured everything for entire plants in the UK and for some foreign plants, though agreements provided for manufacture to take place in all four continents under a variety of different arrangements. "Turnkey" meant a complete package, including preliminary site investigation, raw materials, manufacture, installation, and commissioning of the plant. At the Barrow headquarters, huge industrial cement complexes destined for locations throughout the world were designed and built.

At its peak, the cement division employed around 600 workers—about 70 in design, drafting, and estimating and the rest in the boiler shop, where squads of fitters, boilermakers, and welders fabricated huge machine parts. It was not uncommon for six giant cement kilns to be manufactured at one time in the boiler shop, each 650 feet long and 20 feet in diameter. Started in the 1920s, the division is remembered as an exciting place to work. According to the former technical manager quoted above: "The kilns were done in the boiler shop. I was working with a couple of fitters who did a lot of site work and went out to erect the equipment so they had drawings in their toolboxes of various sites around the world where they'd been . . . really good and I set my sights on joining the cement division in late years. . . . It was a marvelous product with a good world reputation . . . contracts worldwide, it was a tremendous shame some years later when politically they decided to pull out . . . which it was, purely politics."

Various factors that could have precipitated the shutdown were outlined by interviewees, but they agreed that there was a lack of interest in the business by top management and that decisions taken by one individual manager resulted in financial losses on two important UK contracts. These losses were then used to support the decision to close the division. The engineer quoted above, who occupied a senior position within the division, believed that when the division was closed there was still a market for Vickers cement products, and that new business was actually turned away:

The day they packed up the company, cement, was they day they got another complete contract for South Africa. . . . I had occasions in the last 20 years where I could have had orders for cement kilns and cement mills and they wouldn't take them on because they said we hadn't got the capacity in the boiler shop. . . . Just before I finished, about 12 months before [1989], I was at a meeting with the engineering board, where I'd had an inquiry in for mills and kilns and they turned round and told me then, which was to me a ridiculous excuse, which was we haven't got the men with sufficient experience to roll commercial shells—which is ridiculous—if you can roll submarine shells, you can roll commercial shells. . . . They didn't want the job.

This was confirmed by the former technical manager, who said that at the time the closing was announced there was "a contract worth £6 million on the table and various things around the world ready for signing."

But failure to develop and invest during the 1960s also seems to have been part of the weakening of the division. Evans hints at this in his official history, and it is confirmed by the two senior managers. There was also a failure to develop commercial networks. According to the division's chief designer and engineer, "Blue Circle, which used to be Associated Portland Cement Manufacturers, they wanted Vickers to go into business with them, on manufacture and design, because they had a design department, but Vickers wouldn't entertain it."

When the cement division was closed, Vickers sent out a circular to all its customers telling them of the company's withdrawal from the industrial cement market. Yet the rump team left after the shrinkage still received inquiries. About 5 years after the closing of the cement division, a pilot scheme for a method of sewage disposal using incineration in rotary kilns was developed by Vickers engineers in conjunction with one of the former UK water authorities. Building on existing knowledge of cement machinery, this was a pioneering application of the technology. The sewage-disposal method would use its own naturally occurring methane as fuel. But by that time only one specialist in the design of cement equipment was left in the Engineering Design Department at Barrow. Both interviewees asserted that there was a lack of interest by top management in the scheme. According to this respondent, the project was abandoned by Vickers during the final stage of negotiations for a major contract, which eventually was signed with a German engineering company.

At the time of the cement division's closing, there was increasing pressure on capacity in the boiler shop. The Barrow plant had taken on submarine work, naval surface vessels, and the FH70 field howitzer, and it wanted to utilize its boiler shop, where the cement cylinders were made, to fulfill these arms contracts. The cement-plant division had to give way. For the final years of his working life, after the cement-plant division was shut down, the last experienced commercial designer was transferred to armament work, mostly on the FH70—work that, in the interview cited above, he claimed to have loathed for its repetitive nature.

Defense quality control and inspection demands became another impediment to commercial work. Cement-plant design was, in the view of its proponents, more interesting, since it involved site-specific calculations and individually created technical configurations. Defense equipment had to be highly standardized, tolerances were rigidly controlled, and the work was inherently less creative, though heavily inspected. According to a former VSEL manager (interviewed June 30, 1994), some commercial designers who had been moved onto armaments (e.g., submarine shift work) after the closing of the cement division later tried to obtain transfers back to the remnant of the nearly defunct business. "I'd already been in armaments and it was horrible," he said. "The armaments drawing office was just dead, there was nothing happening, they were all hiding, I didn't want to be part of it. . . . They weren't developing anything. . . . Up until I finished I had people that used to work in the cement division asking me if I had a job for them, could I get them transferred from the office they were in to ours, they were all on shift-work submarines, all wanting to come back onto commercial."

Spinoff and Synergy

By 1993 there was virtually no activity within the company that was not directed toward the production of Trident. However, in a vast and technologically complex project full of component technologies, specialist knowledge of materials and industrial processes accumulates in pockets within the organization. Also, although the core business was by then stabilized, designers, draftsmen, engineers, and workers still retained tacit knowledge and broad experience that stretched beyond the contract in hand. In the context of this specialist knowledge and of broader industrial experience, it could be argued that potential synergies were likely to emerge.

As a result of intensive involvement with submarines, considerable expertise in the protection of radar antennas was built up. A radome is a dome that is fitted over a submarine's antenna to provide it with a shield against water pressure or interference. The need for greater protection is implied by the ever increasing powers of detection technology and the requirements of submarines to dive deeper and run faster. VSR3 was the name given to VSEL's own design of a radome, which had a specially cast and mixed form of "syntactic" foam material giving improved protection from hydrostatic pressure and also against detection of the submarine by an enemy.

A leaflet promoting the advantages of the VSR3 (aimed at the export market) was published by the company, briefly describing the technology and relating it throughout to submarine application. The new development would combine "high mechanical strength" to resist depth pressure, "electro-magnetic transparency" to allow minimum distortion of the signal, and "good environmental resistance" to withstand extremes of Arctic and Tropical seawater environments (VSEL Marketing and Customer Services, n.d.). The radome would be made from syntactic material or foam, described as a composite of microspheres contained in a casting resin. Syntactic material was well known and widespread. The VSEL innovation consisted in a new process that allowed larger volumes of the foam to be cast in one operation, "thus permitting the casting of the largest radomes which could be required for use on submarines in the foreseeable future" (ibid.).

Radome "services" (presumably for defense export clients) offered by the company included research and design studies to customers' requirements and quality testing for ability to withstand x rays, pressure, and electrical, environmental, or shock interference. The VSR3 radome was said to improve the electromagnetic transparency (which affected the clarity of the signal) as much as twofold over the existing glass-reinforced plastic (GRP) radome technology. The development appears to have been very successful. Between 1992 and 1993, about twenty radomes were made and sold to a US radar equipment company, Watkins Johnson, to be installed in submarines throughout the world. Although the contracts are said to have been "small" by VSEL's standards (on the order of £100,000–£200,000), profit margins were reportedly large (around 50 percent). A VSEL draftsman and union official, interviewed on March 19, 1993, said: "[On] a small scale it was very successful, there was about 12 of us involved in it, only part-time, it wasn't a full time . . . but it was a very keen group of people, it was something you could really get into . . . you know, the customer was coming over and witnessing pressure testing, you were getting involved with the customer . . . with all aspects. Then we got a huge inquiry from America, from their Federal Aviation Authority."[12] The radomes that were being manufactured up to this point were described as "small," about 12–18 inches in diameter, and the activity was, by VSEL standards, almost like a cottage industry, employing only about a dozen skilled workers. But the US authority was looking at radomes for its civil aviation sites throughout the world, and the size required would have been for units described as 50 feet in diameter, destined for land sites in varying weather conditions from Hawaii to Alaska. "We were looking at a contract, probably 50–100 million pounds," the draftsman said.

Previously, large (civil) radomes had been made from GRP, which was known to be less electrically clear than the syntactic foam material. But the method of mixing and then casting the foam material that had been developed by chemists at VSEL Barrow would allow much larger domes to be made from the superior material while retaining its improved electrical clarity and structural strength.[13] It was only when a large commercial interest in the development was expressed, according to the interviewee, that managers started saying that the product was not "core business." "Essentially I think a lot of our people got extremely worried about it," said the draftsman. "This was a lot of money. . . . It was something they didn't really understand. . . . It wasn't 'core business.' . . . Basically the upshot is that we have sold, or are selling the technology." This draftsman reported that he had been instructed that week to assist in the transfer of drawings and procedures for the technology to representatives of the US company. Because the success of the development itself relied on a process of mixing the syntactic material, he expected that it would take more than transferring documents and drawings to enable successful production in the United States. The innovation relied heavily on the mixing process, and the relevant know-how was held by just a few individuals.

At this time, large-scale redundancies and were being imposed and the company was for the first time publicly espousing a policy of diversification. Consequently, some of those who had been working on the project raised the issue of radome development through their local branch of the staff union, Manufacturing, Science, Finance. When the union asked why the chance for large-scale radome production was being abandoned in a time of job cutting, representatives were told that radomes were not a "core product."

When radome development became "too successful" by attracting high-level interest, it had to be differentiated from core business. What was described as a small, part-time "cottage industry" within the VSEL plant was not a distraction from core business while it remained small. At the component level, here was an example of technical synergy leading to possible spinoff, the development of which was discouraged because of perceived pressure to assist in constructing core business. It may be that the emergence of a new application was seen as a "threat" to production programs, and that this led to a more robust formulation of "core business," whereas a challenge to core business might not have been perceived if radome development had remained less significant.

The "Two Cultures"

For those used to working in commercial engineering, defense work was repetitive, impersonal, and unimaginative. "If the submarine department wanted a valve, well we had most of the standards for these," said the draftsman. "There was no calculation at all, it was just like a purchase order. . . . Same thing over and over." "Milspecs" mushroomed at the expense of design origination, creating different kinds of overheads which militated against maintaining commercial profits. Commercial work did have its advocates among management. One interviewee remembered taking part in management discussions about separating commercial engineering work from defense engineering work by forming an autonomous general engineering unit concentrated in the main production shop where most of the cement equipment and previously the Sulzer engines had been manufactured, but the scheme came to nothing. According to the former manager of the cement division, this organizational separation could have protected commercial contracts from the impact of the spiraling "milspecs" and could have allowed general engineering to survive.

The literature on diversification appears to have two strands of opinion on the effects of defense contracts and the defense "quasi-market" on technical change generally: "crowding out" arguments and "spinoff" arguments. Basically, there is the opinion that arms production and commercial production cannot be run efficiently side by side. Defense work incurs the kind of overheads implied by its special, standardized quality requirements, which work against the central requirement for profit in the commercial market. Therefore, increased defense production can be said to crowd out the more dynamic, varied commercial innovation and production. Perhaps the cement division was crowded out. Alternatively, there is the view that defense research and development has its own form of dynamism, and that it can promote expertise and knowledge, which can then lead to spinoff products with commercial or civil applications. Perhaps radome development could have been carefully managed to allow spinoff. Commitments to finding markets, to promoting new applications, and to engineering—the conditions necessary for development—are essential if spinning off is to work.

There is an interesting divergence in the views of the two former managers, both of whom, while not actively opposing defense contracts, were proponents of continued commercial development at VSEL. One man emphasized the view that defense overheads destroyed commercial production; the other placed much more importance on the role of management:

> There's been a lot of self-destruction going on in the last 3 to 5 years. Five years ago [Vickers, Barrow] had the resources, equipment and labor skills and management skills to take on anything and anybody, given the will to do it. . . . I remember going to a [management] meeting in Bankfield House, 1987ish, where we were discussing what are we going to do to diversify, most of us were saying we used to make Sulzer engines, cement, and so on and what do we currently make? . . . submarine bits and weapons bits, but we used to do them alongside the cement and soap and so on . . . and the production manager starts up "I don't want to, I think this is a complete waste of time, we've got the Trident program, work for the next 20 years. . . . Why bother?" That was the complacency it drove. . . . It was all about management.

A number of critical policy decisions shifted the company away from general engineering. One was taken by Leonard Redshaw, managing director of the Vickers Shipbuilding Group, in the early 1970s. Barrow would pull out of activities such as cement-plant and pump manufacture, and concentrate on the submarine program. Redshaw had earlier championed a lead role for shipbuilding over engineering within Vickers. He had also been one of the architects of the "Buy American" rather than "Develop British" strategy in the 1960s, in negotiations with the United States over the UK's development of a submarine-launched nuclear deterrent (Evans 1978, p. 72).

Barrow still retained the capability to produce such important and innovative technologies as the Constant Speed Generator Drive and the lead flasks for transportation of spent nuclear fuel from the UK's Advanced Gas-Cooled Reactor. Then, in the run-up to the 1986 privatization, Greg Mott, Barrow's chief executive officer, suddenly left his job, reportedly dismissed for opposing the directors' tightening policy of withdrawal by Barrow from every product (even conventional submarines) except Trident.[14] The shipyard was being groomed for privatization and for a single purpose: the Trident program. Its identity was to be solidified, its market was to be structured and streamlined, and its future was to be inextricably linked to the Ministry of Defence.

Hugh Kelly became the manager of the Commercial Department at Vickers in the 1970s, after the closing of the cement division and after the firm's withdrawal from the manufacture of commercial pumping systems and power-generating plants. The department was to handle only spares for systems already installed by the company. He believed that the company lost money and lost tenders for commercial work because its commercial viability was destroyed by its defense-oriented overheads. During the post-World War II boom in commercial production, Barrow engineers had business autonomy and operated within their own overheads. As this autonomy dissolved and engineering was brought under the shipbuilding organization, overheads were generalized. But these overheads were not produced by, and often were not related to, the commercial work being undertaken; rather, they were imposed from "outside." They were a feature of management. Interviewed on January 13, 1995, Kelly said: "We used to work our estimates out on design work and labor intensity—actual labor hours, the time it takes to make a thing, but after we'd put all these together they used to put a percentage on. I used to go into the Estimating Department and say what the heck are you doing putting all this on and they got quite ragged with me and told me "mind your own business, it's nothing to do with you."

Commercial engineering contracts were often made to bear some of the costs associated with areas of the plant that, although not used by the Commercial Department, nevertheless had to be maintained. It is believed that, had there been the will to attract commercial contracts at this time, experience of the market would have revealed that orders could have been obtained if overheads had been lowered.

Conclusion

As Vickers strengthened its associations with the Ministry of Defence, it failed to develop other associations in the commercial world and it weakened those it had. Examples of this are its failure to go into partnership with Blue Circle in the cement business, with the water authority in sewage disposal, and with the US Federal Aviation Authority in radome development.[15]

Successful technical networks have to be built up and maintained by processes of management (often called "heterogeneous engineering"). The "two cultures" of defense and civil technology are constructs that must policed in order to remain separate.[16] Once constructed, the two cultures can act to obviate each other. If this separation is not maintained, they may intermingle, resulting in hybrid spinoff technologies.

It appears from the documentary and oral evidence that, soon after the closing of the cement division, hybrid technologies were still being developed at Barrow during the ongoing process of polarizing these two cultures, before the polarity had hardened. Engineers' creative expectations, underpinned by their experience of the company as a successful actor in the commercial market, had not yet been extinguished, although important relevant networks had been weakened. The stories the engineers could tell about their work and about the company's direction were multiple and creative, but the multiplicity and the creativity were under attack. The black box of production was still open. Only later would it close, as the mode of ordering around the nuclear submarine "core business" gained ascendancy.

II
Reinforcing the Network

3
Technological Roads Not Taken: The Constant Speed Generator Drive

> The Korean order for three CSGDs are for three Shell tankers. Fifty-five inquiries for CSGDs have been received. The prototype (under test) at SMITE[1] performed extremely well and visitors from various companies and shipowners including MoD(N) have been invited to see the prototype on test. An open day was held for the technical press.
> —*Minutes of Joint Monitoring Committee, Vickers Engineering Board, March 6, 1984*

This extract from the minutes of a Vickers internal committee offers a rare view of one particular innovation process within the organization and of its projected market. What follows is an attempt to uncover what happened to the CSGD project and to situate the events within their wider context. The account given here is based on interviews with designers, developers, and managers and on company and journal literature (see figures 3.1 and 3.2).

After the oil crisis of the 1970s, project manager Peter Murrell recalls, the Swiss engine manufacturer Burmeister and Wain pressed Vickers to come up with a more economical way to generate electrical power on a ship at sea. The CSGD was a ground-breaking innovation that applied existing gearing expertise to new technical configurations designed at Barrow and by a small Vickers-owned company in Wales. A major fuel-saving innovation, it appeared to have a bright commercial future.

Within a large engineering company, commitment to any substantial technological project is likely to mean that other projects have not been prioritized for development. The Trident project would be the largest operation ever taken on at the Barrow shipyard. Some idea of the scale of just one of the four submarine contracts can be gathered from the statement by VSEL's chairman, Lord Chalfont, in a television interview with Julian O'Halloran for *Panorama* (BBC1, March 22, 1993), that the shipyard was "now capable of producing a piece of engineering like the

54   Chapter 3

Figures 3.1
Epicyclic gears. Source: Constant Speed Generator Drive (brochure ETP 3.82, Technical Publications Department, VSEL).

Vanguard, the Trident submarine, which in putting it together is something like putting a man on the moon in terms of advanced concepts and engineering."

This chapter places the story of the CSGD in the context of the Trident project and also in the context of the workers' campaign for alternatives to the nuclear submarine program. In retrospect, it is evident that the negotiations concerning the first Trident submarine contract, the workers' campaign for other technologies, and the development of the CSGD were concurrent events. It would, however, have been unusual for these three activities to have attracted equal attention at the time, or to have been accorded equal "status" (in terms of size, employment, political influence). Within such a large company, it would have been unusual for people at the time to have associated these three entities. The Trident actors, the CSGD actors, and the BAEC actors were operating within different yet overlapping networks. Because they worked largely for the same company, they were mostly in the same geographical community, and to some extent they were in the same economic community. Of course, some of the "nonhuman actors," such as gear teeth and technical knowledge, were common to two or to all three of these networks. For this reason, I will attempt to treat the networks symmetrically and to draw parallels among the three episodes.

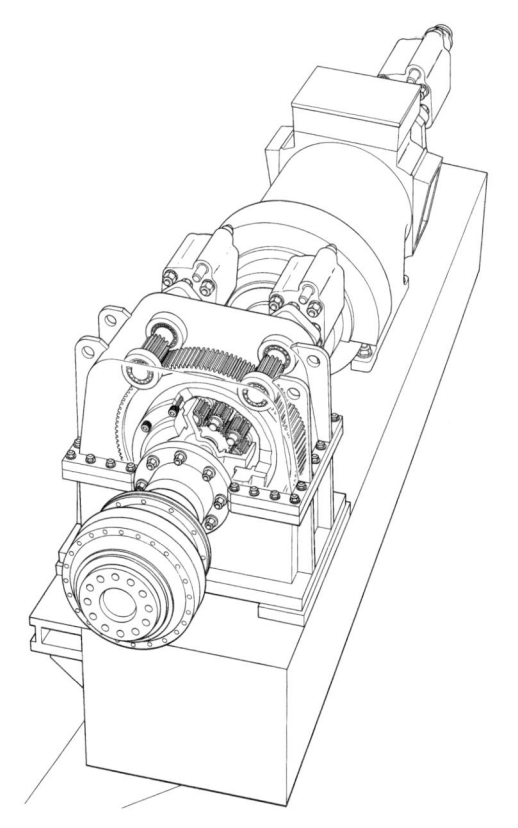

Figure 3.2
Drawing of the CSGD. Source: P. W. Murrell, Shaft Driven Generators for Marine Application (undated report issued by Barrow Engineering Works, probably in 1984).

The development of the CSGD took place during the 9 years between the nationalization of shipbuilding (which formed British Shipbuilders) in 1977 and the privatization of the shipyard (which formed the VSEL Consortium) in 1986. The CSGD was caught between two bureaucracies that had distinct technopolitical agendas. That neither organization saw the CSGD as a priority is an important point that will be revisited. In its report Oceans of Work, the Barrow Alternative Employment Committee criticized the company's failure to exploit its lead in the CSGD market. This fuel-saving gearing technology was precisely the sort of development the BAEC wanted to support in its search for civil, socially useful, but industrially relevant products for VSEL. This was not a kidney machine, a wheelchair, or a washing machine (products that were often associated with workers' defense conversion campaigns or "peacenik" utopias).

The CSGD was at the peak of its development when the first Trident contracts were being negotiated, and some of its designers expressed the view that the innovative gearing project wasn't being pursued vigorously because managers knew that a much larger, more profitable defense project was coming. Other interviewees blamed the pervasive culture within the company of dependence on Ministry of Defence contracts, which fostered a lack of interest in what was seen as a commercial product. Full-scale development of the CSGD would have meant engaging in the commercial market, which would then have become a prominent force in the CSGD network. Also associated with this defense culture are myths about the technical and cultural superiority of defense equipment over commercial products. A former VSEL director, interviewed July 26, 1993, said: "If it didn't fire a missile, they weren't interested." The man who made this comment was the most senior manager who had been associated with the project. All these views, expressed by individuals associated with the CSGD, refer to a time when the company was undergoing complex organizational changes. The technical merits of the CSGD became intertwined with social, political, and organizational aspects of the technology, of other technologies, and of the company.

The CSGD project was, in the early 1980s, the responsibility of the Mechanical Engineering Department (MED) within Vickers Barrow, which reported to the London office of British Shipbuilders. As the shipyard was moving into a virtual monopoly position in submarine building, Vickers had already started to take on advance engineering contracts for the Trident program. Despite the history of rivalry within the Barrow company between engineering and shipbuilding, the CSGD project got off to a flying start, only to be beset by what was officially termed "overspending," with apparently high manufacturing costs and other production problems.

The Technology Described

The CSGD is represented by patent application 8104286 (later 8202043), filed February 11, 1981 at the UK Patent Office and granted in August 1982, which refers to "apparatus for producing a constant rotational speed from a variable speed input." Four inventors are listed: Peter Murrell, John Calverley, Donald Williams, and Douglas James Thomas.

Murrell and Calverley were employed at Vickers in Barrow; Williams and Thomas worked at Compact Orbital Gears (COG), a specialist gear works then owned by Vickers, in Rhayader, Wales.

With a CSGD running off a ship's main engine, electricity could be generated without the need for a secondary power source. In addition, the CSGD could utilize the same lower-grade diesel fuel as the ship's main engine. A CSGD could be retrofitted to an existing ship to replace a conventional secondary diesel generator. Although the initial capital cost of installing a CSGD was about 20 percent higher than that of installing a diesel generator, the cost savings resulting from operating a 700-kilowatt CSGD were estimated at £75,000 per year (£65,000 in fuel savings, £7000 from reduced maintenance and spares, and £3000 from reduced consumption of lubricating oil) (Pringle 1982).

Vickers had built a leading market position in the design and the manufacture of geared transmissions. Promotional literature produced in the early 1980s for the marketing of the CSGD claims expertise with small coaxial industrial gearboxes of varied torque and speed ranges, with marine propulsion gearing for both merchant and naval applications, and with large, high-torque, low-speed gearing for mine winders (Barrow Technical Publications Department n.d.). The Barrow-based gear manufacturing facility of British Shipbuilders was described in technical publications as "vast and diverse," containing precision gear grinding and hobbing machines housed in temperature-and-humidity-controlled shops. The company had developed and held patents on two methods of case hardening for gears: gas carburizing and induction hardening. Three furnaces had been built for case hardening of gear elements by gas carburizing, and the heat-treatment facility had a tooth-by-tooth submerged induction hardening machine.

In 1968, during a period of falling defense orders, Vickers had bought Compact Orbital Gears Ltd. with the aim of developing its industrial gear manufacturing capability. COG had been founded, and was still owned, by Ray Hicks, a gearing specialist who had patented a series of methods of achieving load sharing in epicyclic gearboxes based around his invention, the Hicks Flexible Pin.[2] For Vickers Barrow in the early 1980s, COG represented a facility for manufacturing small, high-quality precision industrial gears. But it was in effect a center of knowledge and excellence for the design and testing of epicyclic gear systems.[3] (An epicyclic gearbox has a central "sun" gearwheel surrounded by a series of smaller rotating planetary wheels around its circumference. This particularly compact configuration allowed optimization of the number of planetary gears within the annulus, with the flexible pin improving the load sharing from all the planets. (See figure 3.1.)

The CSGD used parallel-shaft and epicyclic gear trains to provide efficient transmission. Controlled variation of the rotation of the epicyclics by means of a hydraulic power loop system linking two hydraulic pump units was crucial to maintaining a constant speed. When coupled to the forward end of a ship's low-speed main engine, the CSGD could produce a consistent output as the engine's speed fluctuated. That output could then be used to generate electricity for the ship. According to the literature, the CSGD module eventually developed for use with Sulzer RTA engines would produce 800 kilowatts at 1800 rpm and was capable of producing 1300 kW. The gearbox, the generator, and the hydraulic units would be mounted on a fabricated bed plate secured to the engine by angle brackets. This bed plate would also hold the lubricating systems for the gearbox and the hydraulic systems, and the coolers, pumps, and filters supporting these independent lubrication systems would be secured to the bed plate. The minutes of Vickers's internal monitoring committee stated that, although most of the CSGD's parts would have to be manufactured at COG in Wales, the bed plate would be fabricated at Barrow.

According to the minutes of the Vickers Joint Monitoring Committee,[4] the first CSGD sale was a major order, "gained in the face of fierce competition" in December 1983, from the Korean firm Hyundai for installation in three new oil tankers. The tankers were being built by Hyundai for Shell International, but the order for the CSGDs was officially placed by Hyundai. Vickers (Barrow) had a long-standing relationship with Shell International, for which it had carried out contracts in the past. It is believed that this established commercial relationship eventually helped Vickers to secure the order for the CSGDs against competition from Renk, a German firm that was developing its own fuel-saving marine generator.

The CSGD's developers began to realize that the world market was potentially huge. After the sale of the first three CSGDs to Korea, VSEL's Mechanical Engineering Department reported to the monitoring committee: "With regard to the Constant Speed Generator Drive project, 90 inquiries have been received and MED are endeavoring to produce a technical brochure giving the technical possibilities of the CSGD's for diesel drives and turbine drives. Orders for six units are planned for the coming year with possibilities of 10 units per year thereafter. The total world market is estimated at 250 units per year." (Joint Monitoring Committee minutes, VEB, May 8, 1984)

By August 21, 1984, the Monitoring Committee had been told that inquiries had risen to 111, of which 12 said to be "particularly active." Price estimates were about £250,000 per unit. Six months later, the company had abandoned the entire project and put the Welsh gear works up for sale.

The Evolution of the CSGD

Because the CSGD technology relied heavily on epicyclic gear trains, a specialty built up at the Welsh plant, it was useful to go back and examine Vickers's acquisition of COG, its patents, and its knowledge. This also provided insight into some of the social and cultural tensions between Barrow and Rhayader, between Vickers Barrow and British Shipbuilders, and between production for defense and production for commercial markets. This story also illustrates the difficulties that are inherent in transferring technical know-how. The accounts given by the Welsh engineers indicated that difficulties with transferring knowledge and expertise between the two centers (and later between COG and the Niigata Converter Company) may have resulted in what became seen as "technical problems."

Operating out of a derelict cinema above which he lived, Ray Hicks—a man fascinated with small gears—founded Compact Orbital Gears in 1964. The company soon built up a series of patents and a network of customers. By 1968 it had a workforce of 70 and had moved into a new factory (leased from Welsh Industrial Estates) on the outskirts of Rhayader. COG's first contract for the Royal Navy was for some design study work for a large gearbox ultimately intended for Polaris submarines. In an interview conducted on November 25, 1993, Hicks said: "So we did this design study for a main propulsion warship gearbox and the Admiralty said well, with all due respect, you're only 60 or 70 people strong, you haven't got the resources or the facilities to actually progress this to hardware, you need a big brother." At the time, there were three main builders of warship gearboxes in the United Kingdom: the Vickers Engineering Works at Barrow, David Brown at Huddersfield, and AEI at Rugby. It was Vickers that responded first to the Admiralty's proposal. According to Hicks, Vickers was not interested in entering into license agreements with him for the use of COG patents; it proposed to buy them and the factory outright. Tempted by the comparatively huge resources and potential investment for the small plant that would accrue from having such a prominent "parent," Hicks decided to sell.

Within 3 months of the takeover, Vickers was carrying out a major nationwide restructuring, perceived by many to be part of a "power struggle" between Vickers's engineering and shipbuilding groups. Initially COG had been taken into the Barrow Engineering Works, but it was then decided to move the Barrow Engineering Works (then the biggest and most profitable engineering unit of the global Vickers empire) as an entity into the Vickers Shipbuilding Group (Evans 1978, p. 148). Though still profitable and significant, the engineering section was subsumed within shipbuilding. (This was to happen again in 1977.) Hicks was made general manager at COG, reporting to Vickers Ltd., was given a seat on the board of the parent company, and was retained as a consultant on epicyclic gearing to the Vickers Shipbuilding Group.

As Ray Hicks recalled in the aforementioned interview, there was soon tension between the Welsh and the Barrovian gearing experts over the application of epicyclics:

There was a lot of in-house design and experience at the [Barrow Engineering Works] for warship gearing and they weren't prepared to listen to my general advice . . . so when they made the SSBN (Polaris) . . . they made a pair of gears and set this up to test it on full load. . . . They were very late in their program and despite the fact that I had warned them in advance that I thought the way they were designing it was wrong and it would be unacceptably noisy for a submarine application. . . . By the time they got it up and running it *was* unacceptably noisy, they just didn't have the time left to rectify it, so the Ministry just abandoned the whole project and they reverted to their conventional gearing arrangement of a lock-train power shaft set-up.

An unsuccessful attempt was made at Barrow to develop a range of industrial epicyclic gears. "I again advised them that their choice of application and range was ill advised," Hicks recalled, "but they nevertheless proceeded . . . and to the best of my knowledge they still have 50 gearboxes on the shelf that they never sold." The perception in Wales was that the head office was wasting money on misdirected epicyclic projects—first noisy submarine gearboxes, then industrial gearboxes, and later a range of medium-speed gearboxes (developed, said Hicks, "at a time when the whole of the mercantile marine industry was leaving Europe to go to the Far East"). "There was," Hicks recalled, "a whole lot of abortive spending on that. . . . So we weren't very popular . . . they blamed me for persuading them that epicyclic gears were a good thing . . . and I got the backwash of all this and was pushed into making a whole lot of different industrial gearboxes [at COG], selling them off the drawing board all over the world, Australia, Canada, Siberia." "Selling off the drawing board" meant making the gearbox and sending it straight into service without any research or development. This was how COG was to operate over the next few years. From the viewpoint of the Welsh company, selling out to Vickers began to look like a mistake. "We had," said Hicks, "the worst of both worlds, instead of getting all the resources of a big R&D department and the backup we needed for development work, all that money was soaked up at Barrow so there was none left for us, and we were then left with a remotely located, poorly equipped factory in the middle of Wales, trying to sell gearboxes which were at the forefront of gearing technology . . . even at that time we were selling gearboxes of 100,000 revs per minute."

It was during the 1970s that COG made some variable-ratio gears that were partial forerunners of the CSGD concept. These were hybrid designs comprising an epicyclic system and a variable-speed drive. (Hicks had made his first variable-ratio gear in 1965, before Vickers took over his company). Hicks's first applications for this concept were to hovercraft and pumping systems. Frustrated with Vickers, Hicks left the group just before the nationalization that formed British Shipbuilders in 1977. He started another company, Hicks Transmissions, and continued to patent variable-speed gearing systems. COG spent the next 8 years making spares and providing warship support services, doing general machining and even some non-gearing work for Barrow.

First Signs of the CSGD

In the 1970s, when Vickers was experiencing a contraction of naval work, it decided to look into the possibility of making industrial gears using the epicyclic principle and the Hicks Flexible Pin. Peter Murrell was sent on a marketing trip to visit the European engine manufacturers and learned that in response to the 1970s oil crisis there was a demand for a generator capable of running on low-grade oil. Burmeister and Wain, about to launch a new range of marine engines, asked if Vickers could develop a series of generators to run specifically with those engines. Returning to England, Murrell recalled having difficulties "selling" the idea—difficulties that had to do with Barrow's membership in the warship-building group of British Shipbuilders. According to Murrell, British Shipbuilders' headquarters management frowned on the project because it had no direct application to warships. Murrell, interviewed on March 28, 1995, recalled: "We said well, we're the only gear manufacturer within British Shipbuilders, we're the only ones who would understand how to actually create this product, so what does it matter? . . . We were still an engineering group."

British Shipbuilders then allowed Barrow to initiate a development program for what was to become the CSGD but refused to tie the company to any one engine manufacturer, preferring to keep the field open so as to be able to apply the product to engines of any design. But there were effectively only two major European engine makers: Burmeister and Wain in Switzerland and Sulzer in Germany. Murrell was then dispatched to Sulzer to outline the concept and gauge interest. As he recalled it, Sulzer appeared unsure about the development, British Shipbuilders was slow to deliver the promised support, and meanwhile Burmeister and Wain became impatient and took the concept to the German gear maker Renk, which began developing its own design.

A UK patent application was logged, with an option for European, US, and world rights. Barrow had asked two design engineers, Jim Thomas and Don Williams of COG, to be involved in design and development work on incorporating epicyclic gears and the Hicks Flexible Pin into the new generator unit. Hicks (who had already left COG/Vickers) had no formal role in the development but recalled occasionally giving advice unofficially, by telephone, to former colleagues. Thomas became chief designer and Williams technical manager of the CSGD project, the most ambitious development and potentially the largest venture that had ever been taken on at the Welsh plant.

The Development Process

You see Barrow didn't even believe us when we told them the system would work, so they paid us almost £2,000,000 to build a prototype to prove to them it would.
—*interview with Jim Thomas and Don Williams, November 25, 1993*

The design drawings for the CSGD have been carefully preserved at COG in Rhayader.[5] Mention of the project aroused some bitterness about what the COG designers believe was a wasted development and fruitless use of their expertise by "headquarters." There was resentment that the only outcome of all the design effort and skill application in creating a successful innovation was the eventual sale of the license to manufacture CSGDs to a Japanese firm, the Niigata Converter Company (NICO). By that time, other European competitors had caught up technically and developed their own forms of CSGD (under different titles) in spite of efforts by Vickers to use its UK patent to prevent this.

The gearing designer John Calverley,[6] who worked at Barrow on the CSGD, maintained in an interview conducted on July 26, 1993 that there was nothing "technically" wrong with the Vickers system:

> We turned the Hicks system round, with the CSGD the engine was variable but the output was constant. We designed the CSGD gearbox, breaking new ground, nobody had done it, this was the attraction, nobody else seemed to provide it. We did a couple of prototypes and there were a few teething troubles to do with controlling the angle of the swashplate. And with the one we installed for Hyundai in Korea, initially the speed wasn't within specified limits, the control system was letting it hunt a bit, but it was just the setting, it had to be fine tuned. The SMITE lads and some COG people went out and sorted out the problem. It functioned satisfactorily after that. There wasn't a technical problem, I think it was the finance end. . . . Then the decision was made to run it all down and concentrate on MoD work.

The CSGD prototype tests were indeed a great technical success. The output speed obtained was even more consistent than the designers had hoped for. An electric pickup was used to indicate the variation in engine speed in the input side, which created the signal to indicate that a speed change was needed through the hydraulic pump system. CSGD project manager Peter Murrell recalled on July 28, 1995: "It worked a treat, because we were expecting that we would see some degree of fluctuation, because of the lapse in the electronic system, but it was remarkably flat. . . . In terms of the output consistency the Vickers unit had advantages over the others, the *Marine Engineers' Review* later did a comparison."

The "Official" Company Perspective

A picture of the developing CSGD can be built up from the entries in the Vickers Monitoring Committee minutes. This was a consultative forum, set up under nationalization by British Shipbuilders, in which managers of the various departments of Vickers (Mechanical Engineering, Commercial, Armament, Manufacturing, Marine) met regularly with representatives of trade unions to discuss the progress of current contracts and to address production problems. More a technical briefing body than a negotiating forum, the committee was abolished before the 1986 privatization.

The development of the CSGD coincided with the advance and setup work for the Trident program, first mentioned in the minutes for February 2, 1983: "Progress continues with main machinery and models (design work). Development of design, drawing control systems, planning and budgetary control are all being progressed under Project Orders." Insofar as the order for the *Vanguard* was not placed until May 1986, this gives some idea of the long lead times associated with contracts for nuclear-powered-and-armed submarines. A start was also imminent at this time on the Type 2400 Patrol Class (conventional) submarine, and construction work on two conventional nuclear powered submarines was underway. But while the shipyard was humming with naval contracts, the Mechanical Engineering Department was losing tenders as a result of its "uncompetitive" prices, which were required to include "production costs, overheads and profit." In a competition for wind turbine gearboxes, Vickers's price had been $200,000 higher than "an American company's price of $900,000"; however, "the department did have a £3 million worth of orders for Trident submarine mechanical work on main machinery and noise engineering" (minutes of Joint Monitoring Committee, Vickers Engineering Board, April 26, 1983).

At its May meeting, the committee heard that because British Shipbuilders was incurring losses nationally there could be little hope of investment in replacement machinery locally, even though the Barrow shipyard remained profitable. It was in this difficult financial context that the committee, in its June meeting, first heard about the CSGD, and also about the tensions between producing for the naval and commercial markets: "The CSGD is a small gearbox with good potential, but again, the estimated VEB price for bed plates is twice that of other suppliers. . . . Mr XX said he'd like to point out that the reason why Mr YY kept emphasizing the importance of costs overheads etc., was because he dealt with the commercial market." (JMC minutes, June 21, 1983) It was agreed that a team would be set up to investigate why the CSGD should be incurring such high costs.

It was then reported that, because of the amount of money lost on commercial products, the Mechanical Engineering Department was being especially cautious about accepting orders that year. By July the bed plate for the CSGD had been produced at Barrow and was being sent to COG. The Mechanical Engineering Department was then handling 14 inquiries for CSGDs. But there was a warning: "VSEL has been established as the world wide leader in advanced propulsion gearing design and if this capability is to be realized as orders, more advanced machine tools are needed." (JMC minutes, July 26, 1983)

August 1983 saw the CSGD on target for prototype tests to be run in COG and SMITE in September. At the same time, a £1.5 million order was placed with the Mechanical Engineering Department for Trident emergency coolers, and tenders were being submitted for Trident missile tubes. The problem of investment was raised again when a representative of the Amalgamated Union of Engineering Workers commented that there were machines still in use in the general machine shop that had been built in 1936 (JMC minutes, September 27, 1983).

The Mechanical Engineering Department's December announcement that a Korean shipbuilder had ordered three CSGDs was accompanied by a warning that parts manufactured in Barrow would be too expensive: "The market for the CSGDs is large but because the price is much too high for Barrow manufacture, only a small part of the work can be done at Barrow, virtually no manufacturing parts, assembly and test. The need to reduce man hours, delivery times and prices, to enable the work to be done at Barrow is obvious." (JMC minutes, December 13, 1983)

By March 1984, in the face of fierce competition from the United States, the Marine Department had won a £40 million contract to build the Trident missile tubes for the UK's submarines. This order, to be shared between Vickers and Babcock Power, covered 48 tubes, required almost a million man-hours, and was estimated to last about 7 years. But the Mechanical Engineering Department's committee report was signaling disaster for Vickers's performance in the commercial market. Out of £17.3 million worth of tenders submitted, all the commercial ones had been lost as a result of inflated prices. Only one tender classed as commercial had been successful: a contract had been signed with the Ministry of Defence for a decontamination plant worth £0.75 million. Yet, in contrast with this gloomy performance, the Mechanical Engineering Department reported success with one project (the CSGD), although it emphasized that the units would be in competition for a world market that was potentially very large but in which minimum price had to be the sales criterion. For this reason, it was stated that most of the work on the project had to go to COG, where manufacturing costs were lower because COG was less burdened with defense overheads.

The Mechanical Engineering Department was also negotiating with the US firm General Electric for the transfer of the induction hardening process that Vickers had patented in July 1983, which was then being licensed to GE. It was an example of what Jim Thomas would later see happening to the CSGD. Both knowledge and employment were marginalized: "A license is making money without making the product, it went straight into the Barrow purse. It's profit without workers—that's OK provided you are fully employed, then your licensing fee goes to pay wages and overheads, but they should have seen the writing on the wall." (interview with Jim Thomas and Don Williams, November 26, 1993)

Commercial contracting had deteriorated by May 1984, when the Mechanical Engineering Department's tender for a large magnet frame for CERN (the European Laboratory for Particle Physics) turned out to have been 3 times that of the nearest competitor and that the cost of merely compiling the tenders ranged from £5000 to £15,000. In contrast, the number of inquiries for the CSGD had now reached 90, and the Mechanical Engineering Department was working on a technical brochure outlining the possible applications of that device to diesel and turbine drives. In July it was reported that inquiries for commercial products other than the CSGD had dropped considerably: "Trials of the prototype were highly successful. The first unit for Hyundai is a little late to program but cooperation between all departments concerned is good and it is hoped the assembled unit will be delivered to Kincaid's[7] for the test program early in September. Inquiries for further units of similar power and configuration are being dealt with and the probability of an order for at least eight off very soon, for South African Marine and Mobil Oil is very good. As always price is the major criteria." (JMC minutes, July 3, 1984)

Here was the first sign of problems with CSGD production. The Mechanical Engineering Department had learned that the full-power test of the CSGD could not go ahead at Kincaid's in Glasgow, and an alternative site was being sought. (Unit testing was finally carried out at the Harland and Wolff shipyard in Belfast with the CSGD unit attached to a Burmeister and Wain engine). Meanwhile, Hyundai had asked for the program to be accelerated, insisting that the first and third units be built by Vickers/COG but that the second be built under license by NICO in Japan. Therefore, any delay in production or testing would be serious. Inquiries from potential customers had jumped to 111.

The JMC minutes of August 21, 1984 also record increasing pressure on machinery testing time in Barrow and on general engineering production: "The Trident raft and gearbox is presently on program. The delivery of the CSGD to SMITE has been rescheduled, and this leaves no slack whatsoever, which means that all manufacturing operations must be carried out promptly."

The last available JMC minutes (October 2, 1984) show that there were still problems with production costs at Barrow. Arrangements had been made for testing of the first CSGD unit, which was to be conducted that month and in November: "The two bed plates for this unit have been going through the [Barrow] workshops, and unfortunately man hours are a problem (mostly in the machine shops), 13,000 have been spent. Hopefully the reason for the large number of man hours can be determined. The high overspend for the CSGDs is a real problem when trying to sell such products."

"Knife and Forking"

When interviewed on June 30, 1994, a former VSEL technical manager who had been closely connected with the Mechanical Engineering Department cast some light on the reasons behind the "high overspend." At the time in question, this man was the Machines Manager in the Manufacturing Division. He was in charge of all the workshops at the time when the Barrow components for the first two CSGDs were being made. Prototype work and first design products, especially on a complex unit such as the CSGD, was always more expensive, he explained, than when the technology had settled into production and technical problems were resolved. As he recalled it, "Barrow fell for the old trick of trying to push it into production before the drawings were finished . . . there were a lot of changes going on and that attracts a lot of cost. . . . Barrow was what we call 'knife and forking it,' it's when you're doing something that isn't finished and you're cutting corners to get jobs made . . . working with half finished drawings, it's always expensive. . . . [NICO] waited till we'd debugged it and then launched it. . . . Barrow got labeled as expensive."

Subsequent units could have been made at lower cost if Vickers had retained production and manufactured the system after drawings had been "sealed," workshops properly tooled up, and man-hours better planned. But a financial decision on labor and initial costs was taken, comparing the available prices of the two Vickers units with the NICO unit. This decision, however, was underpinned by what the aforementioned manager described as a policy decision by the directors of Vickers/British Shipbuilders to pull out of all non-defense work.

The Perspective from Wales

Jim Thomas (interviewed November 26, 1993) echoed Ray Hicks on the subject of the poor technical cooperation between Vickers and COG: "They tried to put flexible gears in submarines without taking our advice. But XX always had great faith in COG, he knew that the gearing expertise was here. XX was a good designer, we're all individuals here and we do things very quickly, well XX was like us, he could think that way. He was hived off by himself to keep him quiet until he retired, he didn't conform to the Barrow way of thinking. There was a different attitude there."

The social and technical gulf between the two plants was exemplified by what happened over the CSGD patent application. The Welsh design engineers believed they were not informed that a patent application was being made. However, when the application was published, they saw the papers and complained to Vickers (which later had their names inserted). Thomas recalled: "I feel bitter towards Barrow over what they did over the CSGD and that they probably didn't even acknowledge that we designed it."

According to Thomas and Williams, the reason Hyundai insisted that one of the CSGD units be constructed by NICO was that Hyundai was not confident that Barrow would be able to deliver three units on time. Thus, two were in fact made by COG/Vickers and one by NICO.

The problem was that any development that occurred in the CSGD program in Wales had to be fed back to NICO, and that made the Japanese firm's performance look better than it was. According to Williams, NICO and Vickers had had a license agreement for a couple of years under which NICO agreed to stay out of the UK market. By this time, COG was receiving "hundreds" of printouts from Barrow of the inquiries that were pouring in for the CSGD from the mercantile market. Many of these inquiries were, in the opinion of the COG engineers, suitable for the installation of CSGD units. The volume of inquiries was a result of efforts by Vickers to generate interest in the market. Williams recalled: "We went to Sulzer in Switzerland to try to get them interested and to a couple of Italian shipyards of behalf of Vickers, to a marine exhibition in Korea—they had put themselves about enough to generate all these inquiries." The main competitors in the development of CSGD type technology were perceived to be the West German engineering company Renk and the Swiss engine manufacturers Sulzer, each of whose designs were manufactured by engineering companies under license all over the world. Williams was convinced that COG/Vickers had its own CSGD prototype up and running "before Renk put pen to paper." Some journal articles describing the Vickers CSGD had been published before the patent application, and Pringle's "Economic Power Generation at Sea" appeared during the application process.

Thomas and Williams believe that Vickers, by first advertising the technology and then failing to resolve production problems, allowed its competitors to gain valuable ground: "What happened was that [the competitors] started to make their own, and then where they were selling an engine to a shipbuilder, it was going complete with a CSGD, cutting Vickers out of the market. So [Vickers] started concentrating on retrofitting, but people were just going back to the main engine manufacturer." The Welsh designers believed that Barrow became distracted away from the CSGD project. Williams recalled: "XX went up to Barrow and they appeared to be putting a lot of money into the place. They were sorting out the Trident gearbox and that would have been a big contract. With huge multi-million contracts like Trident, they didn't want to know." But Williams also believed that there were important bureaucratic reasons for failing to exploit the early lead in CSGD technology. Vickers was set up for large-scale defense work; COG was not (although the Welsh factory had long been recognized by the Ministry of Defence as a supplier of equipment conforming to military specifications). COG could be quick and flexible where Vickers could not. Without the many levels of management between the shop floor and the decision makers, COG people tended to know more about a whole development than Vickers people. It wasn't unheard of for Vickers managers to ask COG managers to "pull a few strings for them" with the Vickers board. In fact, the COG designers had comparatively free mobility and access—they didn't experience problems with security clearance. "We were never barred from anywhere," said Williams. "We got into SMITE, which was top security. No, the problem was they did MoD work and they had to have all the paperwork right, every department was set up to do that and if someone came with an instruction, 'we don't want this to apply, we want this to go through unhindered,' I think there would have been mass suicides."

From the Welsh perspective, it would have been almost impossible, in view of the "bureaucratic nightmare," for Vickers to have produced the CSGD at the market price. It would have taken a complete reversal of practice for the company to have started the calculation from the market price and worked back from there. The larger company had so many departments that it wouldn't have been flexible enough to have gone into major CSGD production without a major reorganization, which might have involved job losses at the time. In this analysis, the bureaucracy makes the product expensive.

Barrow-based engineers also spoke of bureaucratic problems, but they laid more emphasis on British Shipbuilders' policies. British Shipbuilders was still trying to find a non-naval shipbuilder that could take charge of launching and marketing the CSGD. Vickers Barrow was seen as a builder of warships and therefore was not considered suitable. It was decided to market the units through Barclay Curle of Glasgow, a part of the British Shipbuilders engineering group.[8] For this reason, Williams and Thomas of COG began to form the general opinion that Vickers's Gearing Department was being given budgets to make design studies, but when those budgets ran out, there would be a switch to a different project, rather than a push for development. Barrow didn't appear to be as committed to the CSGD as COG. According to Thomas, the rationale appeared to involve keeping designers in work at Barrow between large defense contracts, and perhaps the CSGD was one of the mechanisms for doing just that.

What was undoubtedly common to the COG accounts and the "official" minutes of the Vickers Joint Monitoring Committee was Barrow's inability to lower its prices. Nowhere was there any mention of COG's work being too expensive, yet there was constant anxiety about Barrow's production costs. There was also mounting pressure on Barrow's testing and production timetable. However, the accounts given by the Barrow-based engineers Murrell and Calverley placed more emphasis on the directors' lack of commitment to the CSGD and on the pervasiveness of the defense-contract culture. "Given the right backing and the right environment for the product," said Murrell, "there is no reason why we couldn't have made it at a competitive price, the trouble was the VSEL were wholly organized and structured for MoD contracts." Calverley said: "In those days we were in and out of MoD and commercial at the same time. MoD work always took precedence, you would see the MoD work got done on time and the commercial work took second place. You were assured of profit with the MoD work so you'd do it first, you didn't push for commercial. . . . MoD work was the bird in the hand."

With major CSGD production at Barrow being ruled out because of "high costs," manufacturing capacity was needed. COG was to make most of the parts for the first three CSGDs (although, as it turned out, the second was built in Japan). There was already a problem with the costs of the man-hours being spent at Barrow on the bed plates. However, if the many inquiries had been progressed to firm orders, COG probably would not have had enough capacity to fill them. With inquiries coming in (according to Jim Thomas) from virtually every shipbuilder in the world, space had to be found. The CSGD project may have been seen as too small for Barrow (in the light of the impending Trident contracts) but too large for COG alone. It was decided to investigate whether the Glasgow firm Barclay Curle could step in and manufacture CSGDs if COG couldn't. Jim Thomas went to Glasgow a few times to look into that option. Unfortunately, Williams recalled, the device was completely different from what the "out and out welding people" of Barclay Curle were used to—it was "far too technical for them."

The Technology Gets Transferred

COG's designers were then asked to send their drawings to NICO, the Japanese firm that building the second CSGD for the Hyundai contract. The inventors could feel their innovation slipping away from them. The transfer was not entirely straightforward, according to Jim Thomas: "They had a history of problems with that particular one, what they were doing different to us I don't know, because it should have been identical." Apparently the problem was a difference over production methods. COG was purpose drilling parts to suit the particular component, but NICO was producing all the parts by machine. According to Thomas, that resulted in slight mismatching of holes when it came to assembly. Later it was discovered that no further CSGD units were ever built in Japan.

The Bigger Picture

Within months of the "technology transfer" to NICO, COG was sold. The government of the UK had announced in July 1984 that Vickers would be privatized along with the other warship yards. In the pre-privatization realignments, the construction of "core business" entailed differentiating Vickers's market (seen as Ministry of Defence contracts, plus submarines and armaments for export) from the COG market (seen as small, advanced gearboxes for wide application). With COG sold, Barrow did not pursue further inquiries for the CSGD.

The CSGD was caught up in the rationalization and the streamlining that preceded privatization. Ray Hicks recalled that, under the core business structure, COG was not a part of the VSEL business and therefore should be sold off, and therefore the CSGD was not seen as a VSEL product for the future.

Shell had requested two more CSGDs, but by this time there was no prospect of VSEL's supplying them. Peter Murrell and others negotiated with NICO for it to manufacture CSGDs for Shell under license. A new layout, required by Shell for its subsequent two tankers, was designed at Barrow, and the drawings were then sent to Japan. Apart from supplying spares for the units already installed, VSEL then withdrew from the market.

COG remained a small company but became one of the UK's foremost manufacturers of gearing for wind turbines. Ironically, an earlier proposal for Vickers/COG to form a partnership with the US firm Bendix to enter the wind power market had not been pursued by Vickers. "If it wasn't black and sausage shaped, or it didn't come out of the end of a gun," Hicks commented, "it wasn't perceived as a VSEL product."

VSEL continued to collect license income from the Far East for the manufacture of units containing the Hicks Flexible Pin, which continues to be used extensively in marine propulsion systems, in diesel-powered generators, and in power plants. According to Murrell, NICO did not develop the design skills needed to develop an entire CSGD unit. All five of the original Vickers-designed CSGDs are still at sea.[9]

Professional Journals

The potential significance of the CSGD for marine engineering was highlighted as early as September 1983 by the *Marine Engineers' Review*, which noted that major achievements in this field were few but the Vickers CSGD was an exception. However, its development had been progressing slowly, and it was known that Sulzer had been developing a similar project. Vickers had announced that it would be testing the CSGD in October, and Sulzer's prototype would be tested in November. "Considering that the CSGD might still become an important British first," said an opinion piece in *Marine Engineers' Review*, and "it would certainly be a pity if it were pipped to the post by Sulzer's design."[10]

News articles in the monthly *Naval Architect* throughout 1984 chart the progress of engineering companies in the race to capture the market for constant-speed drive units for ships. But the first contract was awarded to Vickers in March (for the three CSGD units for Hyundai and Shell). In May, the Norwegian company Wartsila announced its prototype, called the CR gear, which it had developed from its paper mill machinery gearing for use in marine drives. (No orders were mentioned.) Next came Sulzer with its Con-Speed Drive in September. Three sales to the Lindo shipyard in Denmark were reported, and other contracts were said to be "close to agreement." In September, Renk launched its RCF (Renk

Constant Frequency); two specific orders were described and sea trials were said to be underway.[11]

The View of the Trade Unions

Danny Pearson was the representative on the Vickers Joint Monitoring Committee of the Technical, Administrative and Supervisory Section.[12] As he recalled it, his repeated questions about the CSGD were brushed aside by the managers on the committee. Eventually he took his complaint to the *North West Evening Mail*, accusing Vickers of casting off a possible job-saving alternative to Trident with its proposal to sell COG. "I feel the direction of policy of the company was to move away from this job because it got in the way of Trident," Pearson told the *Mail*. A Vickers spokesman dismissed this as "rubbish" and stressed that COG operated "in a different market to Vickers." The decision to sell was a part of the restructuring of Vickers, designed to cement an identity around its core business of building nuclear submarines. COG was no longer compatible with operations at Barrow.[13]

Pearson was a job planner: from a sequences of functions, he would estimate what equipment and how many man-hours would be necessary to carry out a contract. In his role as the TASS convenor, he heard expressions of concern from some draftsmen in VSEL's Gearing Department about the way the CSGD project was being handled. One draftsman, interviewed March 19, 1993, recalled it this way:

I think the problem was . . . the reason Vickers didn't take it on is it was decided it wasn't one of our products. . . . What sort of decision is that? . . . This is the sort of thing that has really grieved people like me for years and years. . . . From my point of view it's more difficult because when I started in there (28 years ago) there were two distinct companies, the engineering company and the shipbuilding company and engineering were the profit makers at the time . . . we were involved in all sorts of heavy, general engineering and there was a lot of clever people involved.

Then Mr. Redshaw took over and general engineering was just allowed to disappear and we concentrated solely on submarines and MoD work, and this is the problem we've got now, getting back into the commercial market. . . . You were constantly busy, things were constantly changing, developing, but it was all MoD and submarine aligned, we had these other things coming in like the CSGD.

We had people going all over the world trying to sell the idea (CSGD). . . . Again it's typical of Vickers, you can come up with this sort of thing and everybody's great at the start, but when it comes to the point when you've got to start putting real money in to develop it. . . . Then they don't want to know. . . . "It's not one of our . . . it's not a standard product," you've probably heard the expression from Vickers lately, "core products, core business" as they call it.

Two years later, the Barrow Alternative Employment Committee cited the CSGD as an example of management's failure to diversify and protect employment (BAEC 1987):

The major reasons given by management were uncompetitive pricing and high costs of production. Implicitly, the blame was attached to the workforce but for BAEC, it represents a clear indication of the way management's emphasis on defense work and organizational structures to satisfy MoD quality control requirements adversely affect commercial production. In fact, there is a strong suspicion that, as any large success could have interfered with the production schedules of the Trident program, the management were not prepared to jeopardize relations with the MoD by making the required efforts to ensure the viability of any CSGD contracts.

Conclusions

The CSGD was the biggest mistake that Vickers ever made.
—*Jim Thomas, interviewed November 16, 1993*

The Constant Speed Generator Drive was created because Vickers Barrow, as a shipbuilder with very strong engineering skills, was unusual. The combination of skills that existed at the Barrow plant probably was unique in the UK at the time. The announcement of the sale of the first three Vickers CSGDs early in March 1984 came 37 months after Vickers filed its patent application for the technology. There is general agreement among all those who were involved (and the same agreement is evident in the available literature and technical papers) that Vickers led the field, especially at the outset. But the gap gradually closed. In 1984, Vickers, though still dealing with 111 inquiries in August, was only marginally ahead of its competitors. A decision had to be taken late in 1984 to go for full commitment to the project or to pull out. Vickers's lead was in development and technical know-how. The decision was ostensibly based on cost, but in effect it facilitated the construction of core business. The sale of COG in February 1985 followed from the decision.

The concept of the CSGD turned out to be highly successful. The German gear firm Renk quickly took the lead in the race to produce constant-speed drive units, and eventually Renk came to dominate the world market with its Renk Constant Frequency (RCF) model. CSGD project manager Peter Murrell, interviewed on March 28, 1995, said that it was believed that Renk had installed about 150 units worldwide.

A cluster of issues now seem implicated in the abandonment of the CSGD by Vickers. I do not attempt to place these in any hierarchy of causation, only to show the complexity and contingency of technical change. The CSGD was a technological road not taken. However, if its network had been durable and its context more favorable, it could have been a major development and a "Vickers product."

Here are some of the CSGD's contingencies:

- high cost, partly attributable to Vickers's bureaucracy
- bureaucracy, integral to being a prime contractor to the Ministry of Defence
- Trident, the hegemonic program
- local management's ambivalence toward the CSGD, much of which was to be built in Wales
- differing perceptions of technical expertise tension between COG and Vickers
- management's failure to form a partnership with an engine maker to build an integrated system
- the problematic relationship between civil and defense technology
- privatization of British Shipbuilders
- lack of previous investment by British Shipbuilders in Vickers
- construction of "core business"

The picture that emerged from the company minutes was one of uneasiness and uncertainty on the part of management. Most of the issues discussed above relate to the effects of large and technically demanding yet conservative contracts such as that for Trident, which demand long-term planning and which engender inertia that makes an organization unable to respond to quick changes in the commercial market and unwilling to take risks in the face of competition. Unlike the radome development discussed in the previous chapter, which emerged as potential spinoff from defense work, the CSGD emerged primarily from the commercial market. Faced with shutdown of the development, advocates of the innovation tried, unsuccessfully, to persuade both British Shipbuilders and the newly privatized company VSEL that the CSGD could have a naval application.

In spite of its "technical" merits, the CSGD project foundered. But its story shows that Trident itself was not inevitable, as it had to compete with local innovations, just as mass layoffs later showed that it was not the certain social and economic savior of the local community. It may have guaranteed jobs, but it also destroyed jobs by preventing other technical possibilities. In this way, the development of the CSGD and the BAEC initiative together show something of how Trident production was stabilized. They help show us the ordering, not the order, of technological systems (Law 1994).

Full development and production of the CSGD would have entailed the project's having its own legitimate network within the company—a network in which the commercial market would have been a prominent actor. Such a network could have found itself in competition with the emerging Trident network for expertise, production, and testing capacity. The CSGD can therefore be seen as a casualty of the social and product simplification that was taking place during the construction of the Trident actor network. In organizational terms, it could perhaps be seen as a "reverse salient," a radical rather than conservative technical development that had to be managed and accommodated. Production of the CSGD was perceived at the time, as the radome was, as a threat to the emerging Trident network. The Trident contracts were to be vast undertakings, linking other design and production sites in the United Kingdom and the United States. Alternative avenues of production would have been unwelcome. In his description of large-scale technical systems, Thomas Hughes (1987, p. 52) put it this way: "One of the primary characteristics of the system builder is the ability to construct or to force unity from diversity, centralization in the face of pluralism and coherence from chaos. This construction often involves the destruction of alternative systems."

The CSGD began by enrolling Vickers but was then undermined by a series of other imperatives. Vickers was at first enthusiastic about the project but was then turned away from becoming a world producer of fuel-saving generators by costs, by an emergent notion of core business, and by what it decided would be a clash of markets and cultures if it were to be involved in production of both CSGD and Trident.

Trident was being black boxed during the time of the CSGD's emergence. But within the commercial context, as Andrew Webster (1991) has pointed out, a company, in order to innovate, must audit its areas of expertise, which involves classifying technologies—for example, by product area or by production function. This process is what sets up technological choice. Webster maintains that, in having to exercise this choice, a company is forced to unpack the technological black box (ibid., p. 47).

In the defense context, however, abandoning the CSGD was a part of black boxing production at VSEL. The company was to become the United Kingdom's Trident maker, and therefore other futures were disallowed. VSEL simplified its identity by purging itself of those texts/technologies that could compromise it. And as certain associations (including that with the CSGD engineers and that with the commercial marine market) were severed, other associations (including that with the Ministry of Defence), were strengthened.

4
Constructing a Core Workforce

In Barrow, as in many "company towns," the principal industry is viewed with ambivalence by the community. Feelings of attachment mingle with resentment for a company whose ownership may have changed many times but whose workforce (in spite of hirings and firings) has endured these changes and draws on longer memories. Within this collective memory lies the recognition that the industry that feeds and sustains also threatens with its promotion of a community dependence on ever narrowing forms of production.

To understand the wider significance of the company's privatization for the Trident program, it is necessary first to uncover something of the process of worker enrollment that took place in the run-up to the biggest submarine program ever undertaken in the UK. Increased dependence on defense contracts allowed the company to foster a coherent and homogeneous national and international image: that of the UK's monopoly nuclear submarine builder. In parallel, a new local identity was being created for the shipyard's workforce through the company's privatization. A study of contemporary company literature and local newspapers shows that this new identity centered on the attempt to persuade all workers (and their families) to buy shares in the privatized business. Capitalism and core business had reached the shop floor. The workers' share-option scheme was promoted by a management that was attempting to buy out the yard and to fend off the overtures of a powerful multinational rival, Trafalgar House. Some voices attempted to raise a debate about the consequences for technological diversity and employment security, but through most of the privatization episode these voices were drowned out by stronger ones.

Jewels in the Crown

On July 25, 1984, the Conservative government announced that it would sell off the most successful part of British Shipbuilders, its seven warship yards: Vickers Barrow, Vosper Thorneycroft, Yarrow's, Hall Russell, Brooke Marine, and the two hybrid defense/civil yards: Cammell Laird and Swan Hunter. The whole exercise was to take 20 months, with a deadline for completion of March 31, 1986.

The stated aim of this selective privatization was to leave the merchant sector of the UK's then nationalized shipbuilding industry exposed to the market, unsubsidized by its more profitable defense counterpart. The government was effectively separating the control of naval and non-naval work.[1] Commenting on the announcement, the British Shipbuilders chairman Graham Day noted that it was (then) unusual for an organization to sell off its profitable rather than its unprofitable businesses, and that the sale could not be justified on commercial grounds.[2] It emerged that while the defense yards had made £44 million profit in the previous year, merchant shipbuilding had lost £49 million. The government was earmarking the "jewels in the crown" for sale. Opposition MPs expressed concern that the proposed Trident submarine program might fall into "foreign" hands. Partly to counteract these fears, it was made known that if not enough purchasers were to come forward for the seven yards, management buyouts would be welcomed.

The largest single contribution to British Shipbuilders' overall deficit was a £23.79 million loss incurred as a result of worker "sit-ins" at Cammell Laird on Merseyside, according to Graham Day. He warned that the Merseyside disputes were "sending out clear messages" to future customers and potential purchasers. However, the company's balance sheet also revealed Cammell Laird's redundancy and contraction costs of £232 million and a trading loss of £161 million, of which £100 million related to four oil rig contracts. British Shipbuilders had entered into the offshore business in the hope of offsetting the decline in merchant shipping. The move had been "the single most unfortunate decision taken by the corporation." In a hint at tension between civil and defense production, Day attributed the failure to "differences in design techniques." Bad memories of this venture into offshore work, centered around Cammell Laird, haunted managers of the privatized company VSEL/Cammell Laird, providing a disincentive to pursue offshore work. Profit margins in the warship division had held up because of its cushioned position compared with the merchant sector. These shipyards benefited from a flow of Ministry of Defence orders and research support payments, while cost overruns and delays were tolerated. Yet several of these yards were considered to be more inefficient than the merchant yards.[3]

The privatization of the warship yards was fraught with political as well as economic uncertainties. Investors might be deterred by the long lead times that were typical of defense orders, by the dependence on the Royal Navy as the major customer, and by fear of renationalization by a future Labour government. There would be a new climate of competition after privatization, and the industry would be more vulnerable to cuts in the defense budget. Previous shipyard owners such as the old Vickers company, were still in dispute over their 1977 nationalization compensation award under Labour and would be unlikely to bid again for their old businesses.

Arguably the most attractive sale prospect was the Barrow-in-Furness yard,[4] with its near monopoly of submarine construction and "export prospects" (defense submarine exports later dried up). The most profitable shipyard in the privatization program, Barrow's 1984 trading profit was £21.2 million, up from £18.3 million in 1983. Although many national newspapers at the time stressed Barrow's almost complete dependence on the politically vulnerable Trident program to maintain its position,[5] Trident also represented a potentially full order book and secure profits. In contrast, Cammell Laird was considered a bad buy owing to a poor delivery record and difficult labor relations, in particular the recent wave of sit-ins.[6]

Meanwhile, the national trade unions that had been considering industrial action in opposition to privatization decided against such action. The proposed separation of naval and commercial work created problems for national shipyard solidarity. The official policy of prominent shipbuilding unions such as the General, Municipal and Boilermakers' Trade Union was to oppose privatization. According to the business daily *Lloyd's List*, however, workers in the profitable yards believed they could negotiate better pay and conditions in the private sector, and this was causing friction between naval and merchant groups. Shop stewards called for a conference to hammer out joint policy, but while it was "impossible to agree to privatization in principle," it was "equally difficult to oppose in practice."[7]

Shipyard workers began to find themselves embroiled in issues over potential pay and conditions agreements with possible new owners. In a mass meeting at the Yarrow shipyard in Scotland, 4000 manual workers voted not to support a proposed management buyout. There the workers wanted to be free to negotiate with other potential buyers, companies which included Trafalgar House and GEC. Yarrow's management maintained that the buyout could still go ahead without the participation of the workforce.[8] But within a month the buyout was abandoned after financial backing was withdrawn.[9]

It would seem that the role of the workforce in the initial success or failure of a management buyout was significant (as events at VSEL were to show). The issue of emerging relationships between workforce unions and potential bidders became a focus for dispute. In talks between British Shipbuilders and the Confederation of Shipbuilding and Engineering Unions, the unions' Shipbuilding Negotiating Committee demanded that all bidders be invited to talk to the unions so that different offers could be compared. British Shipbuilders proposed that only the successful bidder should be able to talk to the unions. The SNC then threatened to seek independent talks with bidders. Thus, while maintaining opposition to privatization in principle, unions were indicating readiness to negotiate to protect their interests and strengthen their position in the volatile pre-privatization phase.

In addition to the objections of blue-collar and white-collar workers, the Shipbuilding and Allied Industries Management Association voted at its annual conference to oppose privatization in principle while recognizing its inevitability. The association issued a list of demands aimed at protecting employment and pay under private owners and stated that privatization was being "undertaken for ideological reasons and not in an attempt to create a viable and strategic UK shipbuilding industry."[10]

The sale of Vickers Barrow, valued at £100 million because of its profitability and its full order book, became subject to a process of negotiation, brinkmanship, and horse trading. The main contender (the property, construction, engineering and shipping group Trafalgar House) was reported in the *Mail on Sunday* to be having second thoughts about bidding. Rumors that the controversial Trident program might be scaled down or canceled were casting doubt on the shipyard's sale. The giant £230 million Devonshire Dock Hall, being built at Barrow to accommodate the construction of Trident submarines under cover, was seen as a huge capital liability. Re-equipping the yard to facilitate Trident production would cost £300 million over the next 2 years, and Trafalgar House suggested that the government, not any new owner, should bear that risk.[11]

In a strongly worded comment column, *The Guardian* claimed that the government's aims of privatizing the industry and eroding union power had taken precedence over creating a genuinely competitive and successful UK shipbuilding industry. It alleged that Defence Secretary Michael Heseltine had recently intervened to award a Type 22 frigate contract to Cammell Laird, because the bulk of its workforce had defied the latest strike and agreed to new flexible working practices, even though, the column continued, Swan Hunter or Vosper Thorneycroft could have produced the vessel for £7 million less. Although competitiveness had ostensibly been the central theme of privatization, the government was prepared to abandon this in order to break the unions' control over the industry. According to *The Guardian*, "the Cabinet felt that workers who had crossed the picket lines should be rewarded."[12]

Then came the surprise news that Cammell Laird was to be made a subsidiary of Vickers Barrow and that both would be sold as a package. *Lloyd's List* commented that this was a move to make Cammell Laird more saleable. The managing director at Barrow, Rodney Leach, confessed: "It was a surprise to all of us . . . none of us expected the ministers to make up their minds so quickly." Swiftly hitting a more positive note, Leach declared: "There will be more than enough work for everyone. And the move will give us the most powerful shipbuilding capacity in the world."[13]

There was a general perception that the amalgamation with Cammell Laird had been forced onto Vickers, as the cartoon reproduced here as figure 4.1 suggested. This cartoon appeared on the editorial page of the *North West Evening Mail* after the announcement that Cammell Laird and Vickers were to be sold as one.

On August 13, 1985, the first hint of a possible management buyout for Vickers Barrow came. Leach floated the idea in the *Financial Times*, and it was picked up by the *Mail* later that day. The language used in the *Mail*'s lead story, "Workers may bid for yard," marked the beginning of the local campaign to enroll workers into the Trident program itself and the attempt to forge a new kind of loyalty to and identification with the company. The piece began as follows: "Directors and employees at Vickers could mount a joint bid to buy out the profitable Barrow shipyard. . . . The 12,500-strong workforce emerged as a surprise potential bidder for the jewel in British Shipbuilders' crown when managing director Dr Rodney Leach spoke to a national newspaper." The story went on to say that an in-house bid was only one of a range of options which could not be ruled out and there were no firm proposals at the time. It recapitulated government privatization plans and mentioned Trafalgar House as the only other serious bidder. Reference was then made to complications over the sale involving the £230 million Trident construction hall project and the recent takeover of loss-making Cammell Laird. The final three paragraphs were direct quotes from a Vickers Barrow

84 Chapter 4

Figure 4.1
Cartoon (with image of Rodney Leach) published in *North West Evening Mail,* June 13, 1985.

spokesman. No workers and no union representatives (who learned of their supposed entry into the bidding only through the newspaper columns) were quoted. Yet, according to the *Financial Times,* such a management/employee buyout would probably be the largest in UK history.[14]

There were many possible reasons why the local newspaper's story was constructed in this way. In a "company town" there are all sorts of formal and informal links between the company and the local media. Much as other workers changed jobs, journalists sometimes switched from newspaper work to work in the company's publicity office or for one of its in-house journals. A newspaper report should be fair and accurate, according to both the media industry's own training programs and journalists' professional codes of conduct. Often, if there is not enough time to obtain comments from "both sides," a story might be constructed around one event or point of view, offset by a concluding "unavailable for comment" or "could not be reached as we went to press" sentence. A follow-up piece counterbalancing the first can then be published next day. To conform with good practice (and avoid legal problems), if only one party or interest group can be contacted, the wording of the story should be particularly careful.

In Barrow it was very easy to identify the company with the town and the community. The newspaper naturally identified with its readers, who were in effect the community. The terms "Vickers" and "Vickers workers" were often used interchangeably; sometimes this was consistent with the professional code—"fair and accurate"—and sometimes it was not. On the face of it, such a close cultural identification suggests that there was a high level of worker "enrollment" into the company. The conflation of company and town facilitated the fostering of new local identities, such as (in this case) worker ownership.

The buyout and the offer of shares to workers was one way of making the privatization more palatable to workers. But privatization would also serve management interests. The first hint that privatization would mean changed working practices in the Barrow shipyard came the following month when Leach issued this statement: "Breaking away from British Shipbuilders will give VSEL more freedom to determine its own wage levels, choice of product and investment plans."[15]

The Lazard Prospectus

Lazard Brothers and Company Ltd., as financial advisors to British Shipbuilders, launched the official Vickers Barrow/Cammell Laird sale prospectus on October 10, 1985, more than 15 months after the original selloff announcement.[16] Speculation had been mounting that the impending Trident program was proving to be a stumbling block in the sale.[17] Complaints had been made that the sale document was not being made freely available. On July 8, 1985, in reply to a request by a Member of Parliament to see the Lazard sales prospectus, Industry Minister Norman Lamont is said to have responded that the document was written for the "sophisticated investor," or for those who were "professionally advised," and that it would therefore be wrong if it were to be copied and circulated widely.[18] The document itself warned recipients that by accepting delivery of the memorandum they were agreeing not to copy, reproduce, or distribute it without the prior consent of Lazard.

The Lazard memorandum detailed the physical and geographical resources of the two shipyards and gave a brief history of each. Although most of the important shipbuilding landmarks mentioned were civil vessels, the prospectus stated that since the 1977 nationalization the Barrow yard had been primarily a warship facility. The document, which renamed the company Vickers Shipbuilding and Engineering Ltd., outlined its capacity and its technical capabilities and described its assets in detail. This kind of information appeared alongside a description of a management program to drive down costs and "attack" Cammell Laird's losses, which were attributed to "an under-recovery of labor and overheads."[19] The benefits of combined resources and joint tendering for orders were robustly talked up, as were management efforts to "deriv[e] other synergistic benefits from bringing together VSEL and Cammell Laird in a way that will capitalize on the traditional and combined strengths of the two companies."

The company was seen as a warship specialist: "The VSEL Board believes that the strengths of the Group . . . can be welded into an integrated naval and defense manufacturing capability with a uniquely powerful and coherent 'product-market' strategy that cannot be matched elsewhere in Europe or surpassed anywhere."

The document pointed out that Barrow and Birkenhead Merseyside were the only sites where nuclear submarines had been constructed in the United Kingdom, noting that "VSEL has no UK competitor" for such work. Export hopes were high for the new Upholder Class of diesel-electric submarines.[20] Merchant ships were not considered a serious option, because the market was commercially unattractive and because European Community Intervention Funding subsidies were not available to warship yards. VSEL's status as lead yard for nuclear submarines and its commitment to the Trident project had "restricted the need to develop other markets." However, the company was acknowledged to have a long history of success in civil markets and to have retained "all the necessary expertise to exploit them if desired."

Under the heading "Products," the prospectus stated that VSEL was actively developing products for civil engineering markets, including nuclear waste flasks, non-ferrous castings, and spares for earlier products such as heavy diesel engines for shipping and railways, mine winding machinery, and cement making machinery. On July 22, 1985, the prospectus continued, VSEL held more than 30 full or provisional patents in the UK and overseas. In addition, there were eight license agreements for the manufacture of VSEL-designed products. The most significant of these were said to be for the manufacture of the FH70 field howitzer in Japan in conjunction with Rheinmetall (for which an initial fee of £850,000 was received in 1984) and a "gearing manufacture technology agreement with General Electric of USA" (about which no details were given).

General engineering work appeared to have been played down in the memorandum. Yet among the balance sheets at the end of the memorandum was a 5-year analysis of turnover and trading profits broken down by the four main activities: Warship Building; General Engineering; Ship Repairing, and Other. For the year that ended March 31, 1985, profits from General Engineering were reported as £4523 million and profits from Warship Building as £13,054 million, representing one-fourth of the company's trading profits.

In its analysis of the memorandum, *Lloyd's List* concluded that the merger of the two shipyards had been "forced" and would be damaging to the group's income.[21] Cammell Laird's losses represented nearly a third of Vickers's profits, and a further Cammell Laird loss was expected in the following year.[22]

November 1985 saw the launch of the management-initiated buyout under the slogan "Share in the Action." Letters were sent to all 14,000 employees at both yards, and a press conference was held at Barrow. The *Mail* reported on its front page that the management bid had the backing of British Shipbuilders, and that Vickers understood that the government also favored the buyout rather than a takeover by a third party. Again the phrasing used to describe the initiative emphasized a role (yet to be constructed) for workers and implied that the scheme was somehow going to increase VSEL workers' power and influence. "I believe an employee-led buyout would be the best way for us to ensure our future independence," said Leach. An "employee-led" buyout was the best way of ensuring that the workforce would be fully motivated and that the Ministry of Defence would receive the best value for money. Leach also insisted that management intended for employees to have the largest *group* shareholding. The success of the "Share in the Action" venture would depend on whether the government would officially sanction the management to hold a buyout. If so, then shares would be offered to workers to help finance the deal.[23]

The Buyout

VSEL hired a public relations firm, St James Corporate Communications, to assess workers' opinions and mobilize support for the bid in advance of its formal submission. "St James's initial brief concerns employee relations," reported the advertising industry magazine *Campaign*. "However, a substantial press and political effort is expected in the New Year."[24]

The bid was undoubtedly a gamble. A similar attempt at Yarrow had failed in the face of worker opposition, though other management buyouts had succeeded. (A notable success was that of the National Freight Consortium, with which VSEL had held consultations.) The Vickers deal would be the biggest ever UK management/worker buyout. The company had no sure way of knowing how many workers would buy shares before it submitted its final bid in December 1985. Soundings were taken by the PR firm, while reports circulated that there were to be "soft" loans to help workers buy shares. "It would be nice," Vickers's personnel and employee relations director, Rick Emslie, told the *Evening Mail*, "if we got an average of £200–£300 per employee, but all we expect people to do is give as much as they feel able."[25]

Attempts were also made to enroll the wider Barrow and Merseyside communities into the share option. With just 6 days left before the deadline for bidders to confirm their interest in the sale, the newly formed VSEL Employee Consortium (the body that was working up the buyout proposal) suggested that it might widen the categories of those eligible to participate.[26] Those eligible would now be "1) all employees, apprentices and probationers; 2) wives and husbands; 3) children and grandchildren; 4) brothers and sisters; 5) parents and grandparents; 6) pensioners who formerly worked for VSEL or CL and their widows or widowers; 7) residents of Barrow and Birkenhead electoral districts" (VSEL Employee Consortium 1985). The VSEL Consortium's newly appointed non-executive chairman was the US industrialist David Nicholson, who arrived at this time to look over the plant. Urging workers to invest in the business, Nicholson commented: "It is surprising the average British working chap is prepared to put money on the Tote and the pools but hasn't applied this policy to the stock market." Somewhat puzzlingly, he added that investing in Vickers was not a gamble. Stop-and-go policies caused by ever changing ownership would be ended if share ownership was widened along American lines. Wider share ownership would defuse arguments about who should own industry, Nicholson insisted. Also a non-executive director of a US shipyard,

Nicholson promised to resign his US post if there was ever a conflict of interests with VSEL.[27]

A special telephone line was set up to deal with workers' queries about the scheme. From questions recorded from this "Shareline," an "answer sheet" was compiled. It was sent out to all employees a month later. Personal answers could also be obtained.

On December 3, the deadline for potential purchasers to express interest in buying the shipyards was unexpectedly extended by 7 days. Lazard Brothers would only say that the extension was at the request of "an interested party," but there was a strong hint that this party was the VSEL Consortium—still the only group to have confirmed its interest publicly. Trafalgar House had toured the yard but had still made no statement 3 days before the original deadline. The other potential bidder, GEC, had not yet visited the plant.[28] On the day of the postponement, Cumbria and North Lancashire Conservative MEP Sheila Faith gave her support to the buyout scheme. On a visit to the Barrow shipyard, she enthused: "To really involve people and make them understand exactly what business is all about is a marvelous idea."[29] Trafalgar House then emerged as a seriously interested party. Final bids had to be in by February 25, 1986.

Opinion about the buyout was divided among workers, and there was still no official statement from the yard's trade unions. VSEL employees would often write to the *Evening Mail* about company politics anonymously to avoid potential repercussions at work. During the period being considered here, the Letters Page provided some interesting perspectives. December 10 revealed two conflicting views. "If the workers buy the shipyard," one worker wrote, "they will be responsible for earning their own pay, their own bonuses and pensions, out of their own working profits." Workers' money would not be drained away in redundancy payments or subsidies to other shipyards. The workforce should "put our cash where it belongs, in our own yard, for our own good, for our own future; not only Vickers but Barrow too." The writer equated the future success of the company with the Trident program, which would support jobs, whereas alternatives forms of production would undermine them. Lies had been told about Barrow's part in constructing the Trident system: "We shall NOT have any Trident missiles in the docks or in the yard, just as we never had a real Polaris anywhere near us . . . we can't afford the amateur alternative—we can't ALL build Dinky toys." Here the privatization, the worker share-option scheme, the Barrow Alternative Employment Committee, Trident, and employment are seen as interconnected. Here power, economics, technological imperatives, and production alternatives are seen as inseparable. But another letter differentiated between staff grades and workers, condemning the shares as a bad buy. Workers were poorly paid and discontented, the writer alleged, and successful submarine production was "miraculous under the conditions." White-collar workers (termed "staff" at the yard) were paid for sick days; shop-floor workers were not. Only when pay and bonuses were improved would shares be worth buying, the correspondent added portentously.

The unions broke their silence after the VSEL Consortium admitted that it might seek voluntary redundancies before privatization to improve industrial performance. Shop stewards at mass meetings of boilermakers, coppersmiths, and sheet metal workers said that to support the Consortium bid would be a betrayal. At a time when Barrow wages were claimed to be the lowest in the shipbuilding industry, management was proposing to erode benefits—to scrap one tea break, to withdraw an allowance for hazardous conditions on outside contracts, and to remove certain medical coverage within the works. To support the buyout would be "a blatant betrayal of every principle we stand for," said boilermakers' steward Keith Pearson.[30] Other stewards said that the Consortium did not command the shop-floor support it was claiming, and that workers would not win any more control over the company by becoming shareholders. They would be in a permanent minority power and shareholding position.[31]

A timely order for four diesel-powered Upholder Class submarines on January 3, in the face of competition from Scott Lithgow and Yarrow, strengthened the VSEL Consortium's position. Rodney Leach said that the order provided job security at a critical time for the region. "I truly believe we can be a catalyst for the industrial regeneration of the North West," Leach declared, also claiming that the order showed tacit government support for the Consortium's buyout.[32]

St James Corporate Communications reported that 42 percent of employees who had responded to the telephone and interview poll had expressed interest in buying shares. Other questions were also asked in this survey. On privatization, the survey statistics indicated that only 23 percent of Barrovians and 15 percent of shipyard employees were against share buying. On the Consortium, 83 percent of Barrovians were in favor or neutral. On the advantages of the Consortium, 36 percent thought it would be an incentive to work harder. This survey, carried out during November 1985 exclusively for VSEL, could be criticized on the grounds of the methods, the direction of the questions, the lack of detail about the samples, which sections of the workforce were targeted, and, of course, who funded the exercise. However, on the front page of the *Evening Mail* the findings were presented as "objective" data about workers' views on the buyout.[33]

The campaign to persuade workers to buy shares intensified when Lloyds Merchant Bank announced that city institutions would underwrite the Consortium bid, guaranteeing to fund any shortfall should the share issue to employees and townspeople be undersubscribed. The institutions contacted by Lloyds for the Consortium were also offering to donate some of the shares so that each VSEL employee who bought 500 or more £1 shares would receive 150 more shares free. The Consortium announced that there would be interest-free loans to workers for amounts between £100 and £500 and preferential loans for amounts up to £5000. Political support also gathered. Local Conservative MP Cecil Franks tabled an Early Day Motion welcoming the management buyout.[34]

In an intriguing interview with the *Evening Mail*, Lloyds Merchant Bank official Richard Fortin said that there was a legal requirement for each bid to be accompanied by a statement of the advantages of that bid over rival bids and by a 60-page "sale and purchase" document specifying the terms and conditions of sale. "British Shipbuilders," said Fortin, "would prefer to decide on one sum of money versus another sum of money, but that is not practical in this case. Politics as much as cash will determine the outcome . . . and that is why we submit a statement of the advantages of accepting our bid."[35]

Another sign of the political nature of the decision between bids came 3 days before the deadline when the Shipbuilding and Allied Industries Management Association sent a plea to the Prime Minister Margaret Thatcher not to let the fate of the shipyard rest on money alone. SAIMA feared that if Trafalgar House were to win it would transfer work away from Barrow to its newly acquired Scott Lithgow yard in Scotland. "It appears at the moment that money alone could become effectively the deciding factor in the forthcoming selection of the preferred bidder," said SAIMA's letter (which went on to remind Thatcher of her frequently declared support for employee participation).[36]

Rivalry between the two bidders for trade-union support surfaced on consecutive days in the *Evening Mail*. First, Trafalgar House told a union delegation that it was prepared to make a multi-million-pound investment in the company. The unions received guarantees that existing pension conditions would be protected, but Trafalgar House would not discuss wages, conditions, or redundancies until it submitted a bid. The Consortium, which had said that it would not discuss wages, responded by offering the unions several assurances. It accepted that wages were a problem and that there was a need to consider pay levels elsewhere in the industry. There was agreement to discuss elimination of differences between hourly-paid workers and staff. VSEL committed itself to negotiations as soon as possible after the bids were resolved, and it agreed to no compulsory redundancies. "These differences," the *Evening Mail* pointed out, "have consistently been used as key arguments by people at the yard who have argued against backing the consortium or the current management."[37]

Reportedly, a last-minute wrangle between the Ministry of Defence and the Department of Trade and Industry over the timing of the first Trident contract nearly caused a further 6-week postponement of the deadline for bids. The DTI's main objective was privatization, while the MoD's was the successful negotiation of the contract for the first Trident submarine. It was known that in the negotiations the VSEL Consortium was seeking compensation payments for loss of profits and outlay costs in the event of cancellation of the Trident program and that this was prolonging the negotiations over the initial contract. The MoD wanted the contract terms to be finalized before the yard was sold. Ministers accordingly decided to postpone the deadline for a further 6 weeks, sparking an outcry from the bidders and North West MPs. After urgent talks involving Rodney Leach, who argued that the Consortium would collapse if the sale was deferred,[38] the government hastily rescinded the postponement and restored the (second) deadline. This incurred condemnation and allegations of government incompetence in both houses of Parliament.[39]

On the day before the bids were due, Cecil Franks alleged under parliamentary privilege that Trafalgar House had committed a "disgraceful interference in the privatization program, for purely commercial reasons."[40] A 6-week delay would, he claimed, have given Trafalgar House, with its connections in the city, time to discover the Consortium's bid. He also alleged that Trafalgar House intended, after a certain length of time, to close Cammell Laird and transfer its work to the Scott Lithgow plant, along with some of the work from the Barrow yard. Trafalgar House dismissed these claims as "absolute rubbish" and "totally untrue."

All that can safely be inferred from this last-minute debacle is that it was in the MoD's interest to delay resolving the bids because finalizing the Trident contract conditions was the MoD's priority. This was, however, *not* in the VSEL Consortium's interest, as the *Evening Mail* reported.[41] If the shipyard (technically still owned by British Shipbuilders) had the first Trident contract signed, it would have been worth more financially, and therefore more funds would have had to be raised by the Consortium to finance the buyout.

On the morning the bids were submitted, February 26, there was some unusual lobbying at Downing Street on behalf of the Consortium. Ten-year-old Gerard Corcoran, who had won a prize in a Consortium poster competition, arrived to present a framed copy of his entry, "VSEL: Work For Yourself," to Mrs. Thatcher. The Vickers Works Band, which had traveled with the party to London, played as the boy walked down the street (accompanied by his father, who worked in the yard's Technical Publications Department).[42]

As the two bids were being evaluated by British Shipbuilders and the government, both parties were asked to submit additional estimates for the first Trident contract. Trafalgar House then issued a statement that seemed to threaten the prevailing warship-MoD relationship stressed in the Lazard prospectus. It pledged to "launch VSEL in new markets and new product lines" if it won control of the yard. In a rare press conference, the company declared that it was not "the pirate coming over the hill," as it claimed to have been painted. Under Trafalgar House, the yard would be one of the largest and most important operations in a company with worldwide marketing strength. The Barrow plant was not just a shipbuilder but a high-technology business, a company press officer said.[43]

The next day, petitions signed by thousands of yard workers backing the Consortium were handed to the government. The petitions were devised by pro-Consortium managers to help sway the Department of Trade and Industry. Claims by some workers that they had been pressured to sign were denied by Consortium spokesmen. The petitions had been taken around the yard by foremen and department heads. Two workers approached the *Evening Mail* separately, claiming to have been put under duress to sign. Trafalgar House, it emerged, had been interviewing workers off the premises to determine what their concerns were. Those interviews took place at a local hotel, and £5 per worker was paid for expenses and inconvenience. The results were not released, but Trafalgar House said it was not discouraged by the results.[44]

The drama intensified. Two days later, an *Evening Mail* front-page splash claimed that Trafalgar House was confident it had won the battle. By the following day, everybody knew it had lost the yard. "Vickers at Barrow is to stay in local hands after an historic victory for the management-employee consortium," proclaimed the newspaper. It emerged that about 100 Conservative MPs had urged the government to sell to the Consortium, and that a back-bench revolt could have resulted from acceptance of Trafalgar House's "slightly higher bid."[45] Industry Secretary Paul Channon told the House of Commons that the new VSEL Consortium had agreed to better the terms for producing Trident that had been offered by Vickers Barrow. Although the Trafalgar House bid was acknowledged to be commercially superior for British Shipbuilders, it had included one term that Channon could not disclose but which he would have found "very difficult to accept." "Other interests" beyond the remit of British Shipbuilders had to be considered.[46]

The Labour opposition declared that the sale was an abuse of public money because the clauses about Trident were still only theoretical and a proper valuation of the yard was therefore impossible. But Franks was jubilant:

On behalf of the management, employees, local communities and the whole of the north-west, may I express deep gratitude and pleasure at my Right Hon. Friend's announcement? I am sure he will agree with me that the decision ably and amply demonstrates the Government's confidence in the North West and their firm commitment to the principle of wider share ownership, with the opportunity for employees to share in the fruits of their labour. I assure my Right Hon. Friend that the premier submarine builder of the nation can look forward with great confidence to the 21st century and continue to serve the nation as it had done the past century.[47]

Trafalgar House was immediately stung into revealing the contents of the clause that Channon had found unacceptable. It was no more than a demand for compensation if no Trident contract was placed within 2 years. (MoD contracts routinely include break clauses covering cancellation.) Trafalgar House also said that its bid had been £20 million higher than the Consortium's and that it was considering seeking judicial review.

On hearing the news of the Consortium's victory, its chairman, Sir David Nicholson, told the *Mail* that Vickers built "the best and cheapest submarines in the world" and that "the technology is a national asset that should not be put at risk."[48] He was referring to the clause in the offer that gave the government a "golden share" and limited any single shareholder to a 15 percent holding. "This," he said, "is much more than a successful bid for a management and workers' consortium. It is a turning point in industrial relations in this country." Leach said at a press conference that there would be nothing to prevent election of workers to the VSEL board in the future.[49]

Two months later, the Barrow shipyard's entire workforce walked out on strike.

"Share in the Action"

Although a specially worded formal prospectus was mailed to 12,000 workers, it was always impossible for the Consortium to identify all the individuals who might have been eligible to buy shares under the scheme's wider franchise. Contact would have to be made through publicity, which had already been considerable. In addition, before the announcement of the Consortium's success in being allowed to go ahead with the buyout, each worker had received a booklet explaining the "Share in the Action" program and bearing the logo "VSEL Employee Consortium." This booklet, with its large print and cartoon graphics, was intended "to provide information to employees and pensioners of VSEL and CL." Beginning with a letter from Leach that treated potential shareholders as having one identity and one interest, it promoted the general notion of wide share ownership: "We all want to see the widest possible ownership and among employees and pensioners so that combined, they represent the largest single body of shareholders." The phrase "largest single body of shareholders" conveyed a sense of worker power within the company that closer scrutiny did not bear out. Leach himself was already committed to buying £50,000 worth of shares, and other Consortium directors were making similarly substantial investments. Financial institutions could, in the final round, after the 2-week share-option scheme closed, mop up shares to a 15 percent holding. Each could constitute a "single body," loosely defined. And whereas 6.9 million shares were offered to the communities of Barrow and Birkenhead, a planned listing on the London Stock Exchange in the summer later that year would float 35 million shares.[50]

"Share in the Action" had a certain "Your Country Needs You" feel: institutional investors (banks and pension funds) would "stand behind our bid," but "everyone's backing, employee and pensioner alike, will be needed," and "your backing will also demonstrate to the institutional investors that all employees are committed to the business and have confidence in the future" (VSEL Employee Consortium 1986). Shareholders, the booklet continued, would receive a company Interim Report with summary of progress and trading results, followed by the Annual Report. They would then be able to consider this at the company's annual general meeting, after which they would be able to vote on the dividend, the appointment of auditors, and the election of directors. In the booklet, the act of buying shares is likened to participation in the democratic system of government, and, by implication, playing a responsible part in society: "The role of the Consortium's shareholders will be therefore similar to the role of a parliamentary elector. The shareholders will not have a direct say in the daily running of the business any more than an elector has a direct say in the day to day running of the country. But just as an elector helps choose a government, so the shareholders will help choose the Consortium's board of directors." The parallel is undermined a little by the next sentence, which points out that number of votes a shareholder can cast depends on the number of shares he owns, but that is given as a reason for aiming at a substantial shareholding by employees. Dividends and increased share values would not be automatic, the booklet warned, but would depend on general movements in the stock market, on the performance of the business, "and on your efforts." Thus, the booklet, which had the function of interpreting the Lazard prospectus for a less "sophisticated" audience, cast the worker-shareholder in the role of a responsible citizen with enhanced rights and duties.

Workers would be able to buy and sell freely once a stock exchange listing was obtained later in the year. Until then, a special mechanism would allow shareholders to trade. Sellers would have to apply to the Registrars, who would arrange for a buyer through the Consortium's stockbrokers. The special employee loan schemes set up by the Consortium clearly operated in favor of the worker as small investor. Interest-free loans of between £100 and £500 for the purpose of buying shares could be repaid free of any charge over a year or two through weekly or monthly deductions from pay. The preferential loan scheme involved personal arrangements with three local banks with which favorable terms had been negotiated, the amount that could be borrowed ranging up to £5000. Only the interest would have to be repaid in the first year or two. With the offer of 150 free shares for every worker buying 500 or more added to this, the package was extremely persuasive and attractive. Even those wanting only 200–400 shares would get 50 free. Thus, through payroll purchase, workers who did virtually nothing would have 650 shares. This almost cast the non-purchaser in the active role of having to *opt out of* something, in contrast with the purchaser's having to opt into the scheme. As an enrollment device, this was making the worker-shareholder into a default identity and leaving the onus on the non-share buyer, who was seen as "different."

The Prospectus Sent to Workers and the Community

Things moved fast. Shares went on exclusive local sale on the Monday morning after the Friday on which the government gave the go-ahead for the Consortium buyout. A total of 6.9 million £1 shares were offered to about 200,000 workers, relatives, pensioners, and residents of Barrow and Birkenhead. In a dramatic overnight operation, the Consortium had employed a 40-person team in 12 vehicles to deliver application forms and prospectuses describing "your chance for a share in the action" to 12,000 shipyard workers in Barrow and the surrounding district. The VSEL Consortium had only been able to sign the deal to buy Vickers from British Shipbuilders late on Sunday afternoon. With only 2 weeks allowed for the employee share offer, documents had to be delivered on the first available day. Lloyds Bank set up mobile offices at strategic points around the two shipyards, ready for what the management hoped would be a rush of share-hunting workers.

The Engineer reported from Cammell Laird on the first day of the sale, describing the scene as "a confused atmosphere swinging from downright apathy to outright delight." The journal described the interest-free loans and 150 free shares as "sweeteners," but said that white-collar workers might invest up to £7000 each.[51]

A brochure with a glossy cover bearing the title Your Chance to Share in the Action offered by VSEL and Lloyds Merchant Bank was delivered to all workers as the formal prospectus. The detailed financial information about the company required by the 1985 Companies Act was given at the back of the brochure, while the shares scheme aimed at workers, their families, and the community was detailed in simple question-and-answer form in the first half of the prospectus. This prospectus was aimed specifically at Barrow and Birkenhead workers and the local small investors. Page 1 gave the details of the extra free shares being offered to purchasers and of the "soft" loan schemes. There was a pullout Share Application Booklet drafted "for use by members of VSEL Group employees' families, VSEL Group pensioners and resident of Barrow and Furness and Birkenhead."

On the second day of the shares sale, and just one working day after the Consortium's victory over its rival bidder, industrial conflict between unions and management broke out and divisions between trade unionists surfaced. VSEL management was accused of breaking its pledge to open talks on pay the day after buying the yard. "That was last Friday," said Keith Pearson, the boilermakers' shop steward. "It is now Tuesday and still no meeting nor, to my knowledge, has anything been arranged. Promise number one has been broken." A management statement stressed the link between pay and productivity, but the unions countered that Vickers Barrow had "subsidized" British Shipbuilders for years, and that this proved that the Barrow workers had been earning the profits. Meanwhile, three union officials were accused of putting the company's interests before those of their members by traveling to London to take part in the preparation of a Consortium video on share buying.[52]

"Share Fever"

An intensive campaign of persuasion ensued. A full-page advertisement in the *Evening Mail* of March 11 proclaimed "We've Won!" and explained how to buy shares. This was followed on March 21 by a full-page ad warning, in large bold type, "Don't miss the boat!" On the first day of the share sale, three caravans opened for business in the Barrow shipyard at 8:30 A.M. to catch workers on their way in. Queues were reported. The *Mail* reported: "Shares fever hit Barrow today as more than 300 Vickers employees applied for shares within hours of the issue launch."[53] The Consortium and Lloyds were reported to have pronounced themselves delighted with the start. Comments reported from High Street banks were upbeat. A Barrow Lloyds Bank manager was quoted as saying: "Applications ranged from £100 to, in one case, £10,000. I don't think there will be any problem selling the seven million shares." Prospectuses were said to be going like "hot cakes" at Barclays and "like confetti" at the Royal Bank of Scotland. Customers reportedly had started withdrawing savings to buy shares. A Furness Building Society manager was quoted as follows: "It doesn't worry us that savings will be going out of the society. We have got the funds to cover it and we would like to look to money coming back to us eventually. We are a local society and we believe it is for the good of the town."

Just 4 days before the close of the offer, Chancellor of the Exchequer Nigel Lawson introduced a budget that held more inducements for potential share buyers. The stamp duty on shares transactions was halved to 0.5 percent, thereby attracting workers who might be thinking of selling to make a quick profit. A reduction in the standard rate of income tax and an increase in personal allowances would make the average VSEL worker £2 per week better off. The budget also contained a personal shares equity plan that could give people a tax-free dividend on their shares where the money was re-invested. A cut in interest rates was widely predicted. All these measures were portrayed in the *Evening Mail* as good news for share buyers. In its "Budget Editorial," the *Mail* commented: "A banner inscribed 'Share in the Action' flew in the sky, and this was not the only indication that someone up above was looking after the Vickers consortium."[54]

On the day of the share offer's close (March 24), the *Mail* reported a last-minute rush to buy. Queues were said to have again formed by 8.30 A.M. at the Lloyds' caravans, with people waiting "ankle-deep in snow." But 2 days later the shares story had taken second place on the paper's front page to the growing row over pay. While the House of Commons was being told that 81 percent of VSEL workers had applied for shares, the *Mail* was leading with the joint shop stewards' rejection of the company's pay offer.

Cecil Franks told the House of Commons that 11,400 employees had applied—a "phenomenal response." The *Evening Mail* was quick to pick up the irony: "Industrial trouble threatens Barrow shipyard—just days after thousands of Vickers workers applied to buy shares in their company."[55]

There was a wide difference between the two sides. The unions had demanded a £50 per week raise plus improvements in holiday time[56] and other benefits. The company offered 9.9 percent that year and 4.1 percent the next, and it wanted to dispense with special skill supplements and bonuses, consolidating them into basic pay. Shop Stewards' Council chairman Bill Latham, one of the officials who had been controversially involved in making the video promoting the buying of shares, threatened to recommend industrial action if there was no improvement by the next set of talks.

The share offer was hugely oversubscribed. Applicants had applied for nearly 13 million shares when only 6.9 million were available. A limit had to be set of 1000 shares for employees and 220 for others (e.g., local residents). Local company managers, some of whom had applied for substantial numbers of shares, called on the Consortium to increase the combined employees' share allocation from the planned 25 percent of the total to 35 percent so as to better match the level of the employees' interest. But VSEL defended the right of the large financial institutions and the company directors to retain their large holdings. The institutions and the directors had committed large sums in advance, the prospectus had stated so, and those commitments could not now be altered.[57] In this way, the small investors were made "smaller" than they had applied to be; large investors remained the same size.

The "New Identity," the First Trident Contract, and the Strike

The description offered by Barrow MP Cecil Franks can hardly be disputed. The response was "phenomenal," even in view of all the inducements and the campaign of persuasion. How was 81 percent share ownership among workers (even though it was now revealed to be a minority holding of the total) going to affect industrial relations? Was a fundamental change in the relationship between workers and the company about to take place? Would the potential of higher share dividends for workers really bring about an incentive to work harder—an identification between personal and collective performance at work and an identification between the company's and the individual's financial gain? Or would the VSEL chairman's earlier prophecy that workers would see their shares as a quick flutter be borne out? "It was ridiculous," one VSEL draftsman recalled in an interview (March 19, 1993). "You'd go in there in the morning and blokes who'd usually be reading the *Mirror* or the *Morning Star* were buried in the *Financial Times*! Yeah I bought some shares, most of us did. I sold them pretty soon and did up the bathroom."

Major share trading on the London Stock Exchange was not available until the end of July, so workers had no real idea how much their investment was worth until then. Meanwhile, they were voting on their pay offer, and Leach was negotiating the details of the first Trident contract with the Ministry of Defence,. On May 1 a "unique" deal was struck. The MoD gave VSEL generous compensation clauses in the event of the program's future cancellation. In return VSEL offered to build Trident on terms "25 percent better" than were being tabled before the privatization and the buyout. These penalty compensation clauses would cause problems for any future Labour government, since Labour was already committed to cancellation of Trident. But Labour's defense spokesman, Martin O'Neill, said that his party "would not be held to ransom." Franks said the order was "good for the people of the town who have become involved as shareholders in the company."[58]

However, the day after the £650 million contract for the first submarine was announced, ballot results showed that thousands of manual workers had voted to strike. Two sections of the General, Municipal and Boilermakers Trade Union (the boilermakers and ancillaries and the coppersmiths and sheet-metal workers) had been balloted. The latter section had voted 268 to 58 in support of action. The 3500 boilermakers were said to have been 60 percent in favor of a strike. The coppersmiths' convenor said that the main problems seemed to be the proposed new working practices, loss of allowances, consolidation of bonuses, and proposed fixed holidays—not the company's (improved) 11 percent pay offer.[59]

A week later, nearly 4000 workers walked out and 2800 other manual workers held similar ballots. The *Evening Mail* asked some workers how members of the new "employee consortium" could then go on to mount the biggest strike in 10 years in their company. Interviews revealed anger among workers, many of whom were shareholders, particularly over proposed flexible work practices. Some workers, such as ship slingers, feared that their jobs would simply disappear under the new arrangements. Plumber Harry Bolton was quoted as saying: "I think we are going to lose a heck of a lot of our trade to the fitters and boilermakers. If they have these composite working groups it will mean no chance of promotion in my trade in a yard which is completely fitter orientated."[60] A member of the Electrical, Electronic, Telecommunications and Plumbing Union, Bolton said he intended to vote for a strike, adding that workers had been accepting the company's pay offers without a fight for years and that his family's standard of living had gone downhill. A pro-Consortium welder, Jim Irvine, said that he had bought shares, and had encouraged others to do so, but that he would vote to strike. The directors, said Irvine, had broken their promise to make Vickers the highest-paid yard in the country. Welders once had been well paid, he noted, but their pay had slipped badly. Furthermore, fixed holidays were a backward step, and the company was not doing enough to harmonize the conditions of manual workers and staff employees.

Three days later, 5000 white-collar workers, reputed to contain the biggest group of worker shareholders, voted by 95 percent to strike. Even their own leaders expressed surprise at the margin of the vote. "I did not think we would get the large majority that we did," said Bill Latham, joint staff unions chairman.[61] They were angry about reductions in premium rates for Saturday working, in the night-shift premium, and in the double-shift premium.

Many of the allowances and practices that workers enjoyed had been won over many years of dispute and negotiation. The proposal to remove these caused anger. Principles as well as pay were involved. The promises that appeared to have been made at the height of the pro-Consortium publicity drive later amounted to nothing, in the workers' view. When expectations were not fulfilled, unions confronted managers with two statements the managers had made before the privatization: a concession that there was a "wage problem" and a statement that management would commit itself to progressive elimination of differences between hourly paid workers and staff employees. The company had also said that pay would be improved, although the money had to be earned before it could be paid. "The new board want our people to be the best paid in the industry because they are the best performers," Rodney Leach had said.[62]

The joint manual workers committee issued a leaflet comparing pay at Barrow with pay at other UK shipyards.

Although different sections of workers had different grievances, it was the new working practices, introduced late into the management package shortly after the Consortium emerged as the yard's preferred bidder, that crystallized workers' anger. The new work regime, which seemed to attack their skilled identities, included these new provisions:

- interchangeability of steelwork and outfit trades
- tradesmen to use any hand tool
- skilled tradesmen to join flexible working squads
- some skilled work to be done by ancillaries
- skilled tradesmen to do their own slinging
- a new "best working day" without tea-breaks, overtime rota restrictions, or manning restrictions.[63]

In a story headlined "Countdown to Confrontation," Rodney Leach told the *Evening Mail* that, far from losing status, skilled tradesmen would find their jobs enriched by the new practices. "I think the strike as a weapon in a company that's trying to make itself more democratic must be outmoded," said Leach. "The reason quite simply is that if the company bleeds, the employees bleed too."

VSEL launched a campaign to explain its pay and conditions offer to the workforce. Question-and-answer leaflets were issued to all employees still at work and posted to the homes of all those on strike. The company hoped that talks with national union officers, who could "stand back so that they are in a position to see a bit more clearly the full nature of the package," would "defuse the situation and head off an all-out strike."[64] But the dispute rumbled on, with 6000 manual workers on strike and 5000 white-collar workers pursuing an overtime ban and working to rule. After national representatives of manual workers intervened, a new deal was hammered out that ended the strike. Each side claimed victory. The manual unions said they had obtained more money; management said they had extracted more flexibility. Three weeks later, the unions were accusing management of reneging on parts of the deal, though no further official strike action was taken. After threats and counterthreats, a new offer pre-empted further action by the white-collar workers, who eventually settled toward the end of June 1986. Industrial relations overall had not improved; indeed, they had worsened considerably.

Bonanza Town?

Predictions by stockbrokers that shares bought for £1 each in March would now fetch between £1.25 and £1.65 were made as the company prepared for trading on July 31. A VSEL spokesman advised local shareholders to hold onto their investments. "If lots of people sell, the value will fall," he warned.[65] Two days later, a headline proclaimed "Barrow townsfolk have made a cool £4 million."[66] This was purely theoretical. The share price topped £1.60 on the first day's trading, which meant that a worker who had bought 500 under the interest-free payroll scheme and had received the 150 free shares could make a profit of £357.50. The advice was still to hang on.

In an editorial titled "Bonanza Town," the *Mail* said that Barrovians had made a killing.[67] National investors had obviously been impressed by future prospects and the safeguards offered by the government's cancellation compensation clauses. But there was a sour note as the *Mail* attempted to speak for the wider Barrow community: ". . . the shares bonanza is bound to increase the bad feeling which resulted from the very small allocation of shares made to Furness residents who did not work in Vickers."

When launched, the Consortium had advocated the widest possible share ownership within the community. When the allocation was made, however, only nominal amounts of shares went to non-employees.[68] A similar point could have been made about the comparative shareholding position of employees.

Thatcher's Endorsement

The newly privatized Vickers Shipbuilding and Engineering Ltd., with its worker shareholders, its powerful directors, and its city investors, received powerful recognition when Margaret Thatcher became the first prime minister ever to visit the shipyard. Amid tight security, Thatcher opened the huge white construction hall that would become Barrow's principal landmark and ceremonially laid the keel of the first Trident submarine, which her government had ordered 3 months before. Some workers booed from the catwalks high up on the steel platforms in the submarine hall. Outside the shipyard, 200 anti-nuclear demonstrators chanted "No Trident." At the ceremony, Rodney Leach told Thatcher about the workers' huge response to the share offering, which he described as "our unique demonstration of popular capitalism." The prime minister praised the new consortium: "You are doing wonderfully well. This company is pointing the way to the future. . . . Not only must we have the best equipment, we have to be right up front in technology and we have to be right up front in nuclear technology because if a potential aggressor has that technology, we cannot deter him unless ours is better."[69]

The company's first annual meeting after the buyout would be a test of how this popular capitalism was working in practice. Invitations were sent out to 12,000 Barrovians enclosed with their shareholders' annual reports. The company had to hire an aircraft hangar at Walney Airfield, as this was the only covered venue (outside the shipyard) large enough to accommodate all those who might turn up. The company spent £170,000 having the hangar specially adapted for the day. A giant video screen and two large "overspill marquees" were set up to accommodate as many as 5000 shareholders. It was to be a family occasion, with the Works Band playing and the company's products on exhibit. Free shuttle buses were to run from the center of Barrow to Walney. In the event, the turnout was a disappointing 700. The hangar was only half filled. In his speech, Leach told the audience that any future wage increases would depend on improvements in productivity. Very few workers had sold their shares, Leach later told the *Mail*.[70]

The meeting was uneventful. All the motions concerning re-election of directors and a savings-related share scheme were passed with only a handful of objections. Leach said he was pleased with the attendance.[71] Four years later, the annual meeting had shrunk to an attendance of about 250 and was held in the Vickers Sports Club, "with many worker shareholders having sold up their holdings a much smaller venue is required."[72] And the company had received complaints from shareholders about the presence of press photographers at the meetings. They maintained that the meeting was officially private and pictures should not be taken. The company later decided to ban pictures, although reporters could still attend.[73] This is puzzling. "Private meetings" held by large companies are not normally open to the press. But the Consortium, with its initially very public role, had courted publicity. Banning photographers but still admitting reporters is an unusual situation. It was as if the company still wanted publicity but bowed to shareholders' requests for a certain amount of privacy. Now that they were in such a minority, worker shareholders were perhaps less willing to be seen at a company's annual general meeting than in previous years.

What Happened to the Shares

At the time of the company's first annual general meeting, in August 1986, there were approximately 17,088 shareholders.[74] The vast majority of them were small investors holding between 200 and 650 shares. (The latter was a typical amount, representing 500 shares plus the 150 that were offered free to workers.) The list included addresses on the many streets of small terraced houses in Barrow and Birkenhead where workers had lived for generations. Significantly, the total of 17,088 did not represent a full financial year; it was based on sales from April to mid August.

The list for the first full year, 1987, included approximately 17,280 shareholders. A very large portion of these entries represented *closings* of share accounts, the overwhelming majority of which were those of small holders. In the 1988 register, the total number of shareholders was 9920. This pattern was repeated over the next few years. By 1993, the total number of shareholders was approximately 4000. Again the vast majority of entries represented the closed accounts of small investors who undoubtedly had been attracted by the rising share price.[75] This analysis shows that the majority of workers sold their shares within a year of buying them. Many used the proceeds to take a vacation or to improve their home. In the main, they had seen themselves not as shareholders "owning" a permanent stake in the company but as employees taking an easy opportunity to gain a quick few hundred pounds.

The promise of a say in the running of the company had not materialized. In 1989, as work on the Trident contract began to run down, the company imposed massive redundancies, using criteria rejected by workers. Negotiations between management and unions virtually ceased. Far from operating in a participatory democratic framework (as intimated in the "Share in the Action" literature), life in the company became *less* democratic. However, the price of a share soared from its original £1 to more than £10.90 in 1994 and then to £17.00 in May 1995.[76]

From the perspective of a Barrovian, the imposition of waves of redundancies since 1989 appeared to mirror this climb in share prices. In a single issue of the *Evening Mail*, a redundant ex-worker shareholder was able to read on page 1 about the company's huge Stock Market success, and on page 5 of 600 more job cuts. That day the price of a share had soared to unprecedented 865 pence because, according to a stockbroker, investors were prepared to pay above market value for shares in VSEL. At the same time, the company was cutting its engineering capacity dramatically. The *Evening Mail* reported: "VSEL is to slash its light engineering department in a ruthless bid to cut costs, giving the clearest indication yet of where the jobs axe will fall next."[77] The company claimed that subcontractors could provide simple components more cheaply. It also warned that the foundry and the gear-tooth processing section were up for review.[78] By the end of 1993, the company's profits were becoming almost embarrassing in the local community. Announcing an 11 percent rise in profits for the previous 6 months, which had resulted in a £270.4 million cash "mountain," VSEL's new chief executive, Noel Davies (Leach's replacement), "quick to defend the sum," said that the money represented security for future work at the Barrow yard: "People in Barrow might consider £270 million to be a very large sum of money, but we need to get this into the balance books so the Ministry of Defence can see we are a viable company and able to take on new contracts."[79] The increased profits were due entirely to the Trident contract. Job cuts had not contributed to the company's profit at all, Davies insisted. "It is both very expensive and painful to lay people off. It certainly does not help profits at all during times of crisis." The hard struggle of the past few years was likely to continue, he added.

Conclusion

The viewpoint of this chapter is predominantly local, although different worlds do map onto one another in the story. One can see this by considering the identity of the ex-worker shareholder as a local craftsman—once enrolled into Trident production, into financially supporting the company, and into association with the stock exchange—now jettisoned from all these (related) networks. The identity of the ex-worker shareholder brings together many elements of the privatization story.[80]

Briefly described, the process of "translating" workers so they would become enrolled into the Trident sociotechnical network involved the following activities:

- conflation of workers, the community, and the company
- construction of the Trident as core business that guarantees workers' jobs and "backs Britain"
- association of workers' share ownership with participatory company democracy
- reconstruction of workers as profit makers and share capitalists.

Figure 4.2
Cartoon published in *North West Evening Mail*, November 11, 1993.

When more than 80 percent of workers actually bought shares, this translation process appeared to be a great success. But the 1986 all-out strike suggests that workers held a deep ambivalence about this new role. Like the fishermen of St. Brieuc Bay who defied scientists by suddenly setting out and culling the precious scallop stocks instead of continuing in their new role of conservationists (Callon 1986a), the Barrow workers quickly turned in a direction not foreseen by management. Whereas the company (sometimes assisted by the press) portrayed itself as representing the town and the community, and share buying seemed to tie management and the workforce together, the strike that came hard on the heels of the buyout reflected a decomposition of all these new alliances along traditional social class lines.

Workers voted in overwhelming numbers for a strike, surprising even their own leaders. The 1986 strike was preceded by a series of secret ballots—tangible manifestations of democracy in action. However, the new-style democracy within the company was not trusted. Workers didn't want to wait for the company's annual meeting and didn't believe that it would be a forum for redressing their grievances. Continued ownership of shares was not endorsed by the majority of worker shareholders, who cashed in their shares to meet immediate needs and/or obtain immediate pleasures. Their participation in capitalism was therefore transitory. Capitalism was served by their ready participation, but most of the workers soon reverted from a "bourgeois" to a more "proletarian" identity. Ownership of shares never really signified a common purpose between management and workforce.

What was the outcome of this whole elaborate exercise? Was enrollment served? Rather than describe this privatization process as a *deliberate* strategy designed to secure the successful production of Trident, I will say that worker loyalty and identification with the program were major elements in that success. My overarching impression of the worker share-option scheme is that it was a huge distraction from the political and technological problems inherent in taking on the Trident contracts, and that it muddied the political waters. Workers were divided as to whether to buy shares. Supporting the "local" management rather than the "outside" bidder (Trafalgar House) was an initial incentive to join the scheme for some workers who might otherwise have been unlikely to participate.[81]

Rather than weaken the unions, the aftermath of the buyout actually seemed to have strengthened them initially, to judge by the solidity of the strike. But a massive weakening was soon to follow with the vast redundancy program of the early 1990s. As the largest management buyout in the UK at that time, with certainly the biggest worker share option, the venture aligned with the Thatcherite ideology of share owing and property owning. The prime minister herself looked favorably on the scheme, and she visited the yard as soon as it was completed. This was the new, acceptable way to privatize a company. Enrollment here was partial and was masked deep ambivalence.

III
Alternatives

5
Winning the "Technical" Arguments

The Trident program was sold twice. First, it was sold as crucial for national security to the United Kingdom's public as a whole. Second, in Barrow, the expected four-submarine series was sold as the only project large enough to keep the town in work for the foreseeable future. Simultaneous with this double sell (and with the privatization and advance work on the Trident program), a campaign for alternative production was mounted inside and outside the Barrow shipyard by a group of workers. This chapter considers how this campaign crystallized around "winning the technical arguments."[1]

We have seen how the boundaries between "politics" and "technology" were constantly being negotiated throughout the story of how the shipyard came to be seen as a Trident producer. This process of negotiation was also a fundamental feature of the efforts by the Barrow Alternative Employment Committee to influence the content of production. The campaign aimed to promote the production of technologies that it believed would both sustain jobs and be more "socially useful" than the Cold War edifices with which the shipyard seemed destined to be associated.

A strong theme in the early stages of the Trident-in-production story was the selling of the program to a workforce and a community. The Trident contracts were represented as guaranteeing jobs and prosperity in a town that was by then almost exclusively dependent for its livelihood on defense contracts. The first Trident contract, negotiated with the Ministry of Defence in the early 1980s, was presented as the only shipyard project that could keep the town's workers employed. In that sense, building Trident took on an (economic) inevitability locally even while the program was still politically uncertain in the UK. The project was also portrayed locally as the technological successor to Polaris, with the effect that both its radically enhanced strike capability and its high-cost sophistication or "baroqueness,"[2] were less apparent. Accordingly, in Barrow, Trident's "inevitability" appeared as both economic and technological. Add the experiment with capitalism discussed in the previous chapter and arguably there existed a sort of "deterministic triad" (Heilbroner 1994) of economic determinism, technological determinism, and capitalism.

Building the Trident submarines appeared to conform to a locally constituted "trajectory of production," a widely held perception that helped to black box the production program. Eric Montgomery, a fitter/turner and a former chief union convenor at Vickers Barrow who had a reputation for militancy,[3] had worked on many classes of submarines, from the small Oberon class boats up to Resolution/Polaris and the hunter-killer class. In an interview conducted on June 28, 1993, Montgomery said: "There was no difference, Polaris was just an ordinary sub but with a nuclear propulsion. With Trident it was sheer size, but basically it was just the same . . . with of course tighter safety for the bigger reactor, and more security."

By the time the Trident submarines were proposed, Eric Montgomery no longer worked in the yard.[4] Familiarity with submarine production, and especially with the Polaris program, could easily support the view that Trident was "just another boat," but in the 1980s local opponents of the Trident program (including Montgomery) began to realize that different kinds of arguments were needed to revitalize the debate.

If Trident was presented as both the economic and the technological successor to Polaris, the company appears to have reasoned, "selling" it to the workforce would be less politically awkward. However, because of the sheer scale of the Trident program, workers began to notice that both general engineering activity in the yard and also specialist areas of skill (such as gearing expertise) were disappearing or being scaled down. Production was to be almost entirely devoted to Trident, and this exclusivity began to create a near obsession locally about job security. Precisely because building Trident was to "guarantee" jobs; *not* building Trident, or seeing the project canceled, came to represent mass unemployment. However, for some active trade unionists within the shipyard the argument ran in the opposite direction: undertaking the full Trident program meant that a short-term jobs boom would give way to an employment vacuum in the future, when the contracts were completed. For this group, Barrow's new role as the UK's monopoly "lead" submarine yard[5] was seen as a high-risk company strategy that dangerously courted future impoverishment.

Production within Sociotechnical Networks

The Trident system was a social network that had to be constructed and then kept in place. This network had to be held together during the process of the military, political, and technological transformation of the weapons system in the United States. Thus, the network moved from serving the stated strategy of "ultimate deterrent" exemplified by the Polaris system and the "logic" of mutually assured destruction to serving the "accurate" counterforce-oriented weapon system known as Trident D5.[6] But the sociotechnical stability of the system also had to be achieved at the production/industrial site of the program. In the General Dynamics shipyard in the United States, production and technical problems were requiring enormous feats of technical and social engineering to secure the program's credibility (Tyler 1986). In the United Kingdom, the problems of stabilization were political/strategic nationally and political/technological locally.

Latour (1987) pointed out that the number of scientists and engineers engaged in research and development is minuscule in relation to the size of a country's total workforce. However, the workforce engaged in Trident production has been necessarily large and diverse. In the United States, the missile engineers, scientists, Navy personnel, and politicians associated with Trident routinely presupposed a world in which the artifact would be produced. True, many defense systems never get realized, they may have a largely rhetorical role, and then they may get canceled. But at the end of its many strategic and scientific transformations in the United States, Trident D5 and its submarine were successfully brought into production and deployed, having survived many serious contingencies.

In spite of the perceived production "trajectory" of constructing ever bigger and more sophisticated submarines, the labor of assembling resources and a workforce—that is, machines and people—still had to be undertaken to produce the new system. One of the episodes in which the UK's Trident production network had to be made secure occurred during the growth of what was effectively a counter-network: the Barrow Alternative Employment Committee. By advocating alternative products for the shipyard, the BAEC challenged the role that was being promoted for the company and its then expanding workforce. So here were two actor worlds in the process of construction: that of Trident and that of alternatives. There were, then, two opposing attempts at enrollment going on. The first of these succeeded; the second largely failed, although some of the BAEC's ideas were for a time claimed by others in the aftermath of Trident production.

This challenge or problematization of the prevailing or dominant network provides a case study in the sociology of technological change. A historical account of the formation of the BAEC shows that exclusive defense production, which later did lead to mass structural/technological unemployment, was not inevitable. This case of lost alternatives, of more roads not taken, reveals intimacies in the relationship between technology and power.

The Enrollment of Workers into Trident Production

Actor-networkists use the term "enrollment" to describe how identities for people, institutions, and things (machines) are defined and distributed so that an actor world can be built and maintained. The strategies by which people and machines are controlled have been termed "translations."

Various strategies were used to enroll the Barrow shipyard's workforce into the building of Trident. Once sufficient workers' interests were aligned with the project, they would become essential links in the chain of Trident actors. They would form the furthest (in space) and latest (in time) outreach of a huge US-UK network.[7]

It was often said by Barrow trade unionists opposed to nuclear weapons that local workers were "economic conscripts" in the Trident project.[8] In some ways this was a relatively new situation, according to Eric Montgomery (an assistant to former Barrow Labour MP Albert Booth, who had been defeated in the 1983 general election when the traditionally Labour town elected a Conservative MP for the first time[9]):

During the period from the start of Polaris right through until Booth lost his job . . . 83 election . . . up to 2 or 3 years before that er . . . it was quite easy, relatively easy campaigning against Polaris . . . there wasn't any immediate threat to jobs. But then there was the campaign against Trident, there was the switch . . . you know . . . there was quite a lot of feeling which hadn't been around before in terms of "Trident means jobs and anybody who's trying to knock that is endangering our jobs" . . . a situation which hadn't actually prevailed before . . . the previous election he (Booth) was one of the very few to increase his majority and that was in spite of a campaign in terms of Polaris, but it [Trident] was accentuated around that [later] election . . . and then there was the Falklands War and we'd just come through a recession.

The story of Montgomery's rise and fall within the shipyard unions is described on pp. 52–53 of Beynon and Wainwright's 1979 book on the early Vickers empire and workers' earlier attempts to form a national organization of shop stewards. That account and Kaldor 1976 offer a glimpse of the history of politically precarious attempts by trade unionists to identify alternative civil projects for Barrow and for other facilities.[10] Much of this history pre-dates the higher profile Lucas Aerospace Shop Stewards' Plan, which in turn had an influence on the BAEC. In a sense, Barrow trade unionists in the mid 1980s were reviving a significant (if minority) labor tradition, in their shipyard and beyond, of campaigning for industrial conversion away from defense work.

"Conscripted" by economic pressures or not, Trident workers have been essential to the completion of the Trident system, and without their cooperation present and future defense/industrial networks might have been threatened. This element of economic conscription may present some difficulties for social-constructivist and actor-network approaches to technology. Can the "softer" notion of enrollment expressed by Callon and Latour adequately describe the harshness of economic conscription or the coercive nature of the selective compulsory redundancy policy that later came to be imposed at the shipyard? And what is the role in such networks of the redundant worker or the ex-worker shareholder—former actors in the network who have since been jettisoned? Coercion into networks is not wholly inconsistent with an actor-network understanding of technology. A parallel idea might be Latour's example of people who need flour becoming dependent on the miller and his new technology, the windmill. However good they were at grinding with pestle and mortar, individuals could not compete with the mill (Latour 1987). The shipyard workers campaigning for alternatives could not ultimately compete with Trident as a mass provider of work.[11] Perhaps conscription and coercion are just forms of enrollment at the extreme end of a process in which there are degrees of choice and resistance.

Strategies of enrollment into the company-Trident network should be seen in the context of a long process of deepening dependence of Barrow workers on the Ministry of Defence and the increasing militarization of the yard and its engineering shops. Before the launch of the UK's first nuclear powered submarine, *Dreadnought*, in 1960, Vickers had built all the big passenger liners for P&O and previously for Orient and Canadian Pacific, and the 100,000-ton oil tanker *British Admiral* for BP. Barrow's defense orientation was considerably strengthened by the Polaris program. The cancellation in 1962 of the American Skybolt air-launched ballistic missile, which the UK was to have purchased, was followed by the decision to buy the US Polaris system. Though this highlighted the UK's almost complete strategic dependence on the United States, it was also to signal Barrow's exclusive role as producer of the delivery platform for the newly ascendant naval (over airborne) nuclear "deterrent."

VSEL then became the most defense-dependent prime contractor in the world, with more than 95 percent military production.[12] The baroque nature of the Trident technology shaped the composition of the workforce, its balance of skills, and the resources available to it. From *Dreadnought* through Polaris to Trident, ever more demanding specifications were imposed to maintain safety and accuracy under extreme operating conditions. This began to result in what the BAEC called "technologically sophisticated conservatism." The only industrial "spin-offs" from this defense dependence were the adaptation of marine howitzers for field use and a short-term contract to build nuclear waste transport flasks (BAEC 1987). The Trident program has provided virtually no opportunities for technological synergy, owing to its unique defense characteristics, its baroqueness, the sheer size of the production effort, and the company's and the MoD's preoccupation with secrecy. The BAEC predicted this situation in its report Oceans of Work. This dearth of industrial creativity was also attributable to the MoD's explicit policy of "not having an industrial sponsorship role or a responsibility for technologies unrelated to defense."[13]

How the "Counter-Network" Began: Take the Money, or Open the (Black) Box?

The run-up to and the negotiation of the first UK Trident contract, awarded in May 1986, coincided with intense local and national political activism. The early 1980s saw phenomenal growth in the peace movement in the UK. There were waves of opposition to US military bases, such as the innovative women's protest at Greenham Common and the resistance camps at the Molesworth base (before their eviction by the Ministry of Defence). The fortieth anniversary of the bombing of Hiroshima and Nagasaki attracted widespread media coverage of associated memorials and protests. Around this time were also the long and bitter coal miners' strike, the violent Wapping newspaper dispute, the bombing of Tripoli, the Ethiopian famine, and the bombing of the Conservative Party's Brighton conference by Irish republicans. All these events took place in the run-up to the 1987 general election.

People in Barrow at this time were concerned about whether or not government orders for Trident would materialize and what the Labour

Party's revived unilateralism and its non-nuclear defense policy would mean for local employment. The long-expected privatization of the shipyard in 1986, which gave rise to the creation of worker shareholders, was of major interest to the community. Less obvious publicly, but related to all these issues, was the formation of the Barrow Alternative Employment Committee. The elaborate 1985–86 privatization exercise in Barrow was a distraction from what the BAEC saw as the most important issue facing the shipyard during the run-up to Trident: the content of production. In opposing the privatization, the BAEC was running against the tide of what appeared to be local opinion and a vocal "jam today" trade-union lobby. Whoever bought the company, argued Danny Pearson, TASS convenor and one of the BAEC's founding members, would be concerned only with immediate profits and shareholders and not with long-term job security or investigating new products and markets. Even the old (pre-1977 nationalization) Vickers management would sometimes take on a loss-making contract in order to keep skilled teams together, Pearson recalled in an interview on November 6, 1991.

Black Boxing Production

The story of the Barrow Alternative Employment Committee is partly a story about workers attempting to open the black box of industrial production. How does the production process become black boxed? In workplaces where there is severe division of labor, industrial production can become black boxed by removing from the worker any sense of the whole product, its origins, and its users or the social role created for it. In the case of the Trident submarine worker, who is typically highly skilled, a sense of the whole product is partially retained through the skilled nature of much of the work, through discourse with colleagues, and through the high-profile launch and naming ceremonies (which have always been important features of industrial tradition and community history in shipyard towns).[14]

Trident production became black boxed in a different way: the program was presented as the only technology, the only contracts, the only significant work available. Question producing Trident and you shut down the shipyard. The box had to stay shut. By tracing the history of the BAEC's formation, and by examining its work and the public responses to its reports, we can see that this workers' committee was arguing against Trident as a technological imperative. Part of the BAEC's work had the effect of deconstructing that imperative by breaking a submarine down into its component technologies, skills, and resources and advocating alternative uses for them.

How the BAEC Was Formed

The earliest reference within the scope of this study to the sense of unease among workers about building nuclear-powered and nuclear-armed submarines was made by former TASS convenor Danny Pearson, who noted in an interview conducted on November 6, 1991 that workers had been "quite happy to be turning out 16 and 18 submarines a year during World War II, when we were taking on the Fascists—morally we felt everything was all right." Later generations of submarines were seen, said Pearson, as a "natural step" insofar as "it was only natural that Vickers should get the orders." But the sinking of the Argentine submarine *Belgrano* during the Falklands War in May 1982 was different: "Morally we were just in it up to here." The attacking submarine was one of the few UK boats not actually built in Barrow, but it could easily have been the product of Barrow labor. The controversial sinking, with the loss of 323 lives, caused uneasiness among local people, according to Pearson: "Up till then the submarines we'd been building were only 'models' on NATO exercises, getting moved about. . . . Everybody was quite happy go on producing things that were . . . , if you like, moral deterrents, but not used in a practical sense." After the *Belgrano* sinking, Pearson said, there was disquiet among his union members about the way in which Trident might be used. Unease was expressed not just among left-leaning workers but also among "the more hawkish types, and steady foremen." And outside the yard, the Barrow Campaign for Nuclear Disarmament conducted a street poll on the issue and found similar concerns in the community.

Through the early 1980s, the Barrow Trades Council, at the time a focus for activist and "more progressive" ideas among the town's trade-union movement, met regularly. When the council received a general mailing from Alan Milburn, secretary of the Northern Trade Union Campaign for Nuclear Disarmament, offering to visit a local meeting, the council accepted. The resulting meeting was described as a "receptive" gathering where workers expressed concern about future employment levels in the shipyard and interest in exploring alternatives. Trade Union CND (TUCND) then contacted the regional trade unions about possible defense conversion initiatives. A series of meetings, again attended by delegates from Barrow, were held in the North West. It seemed that Barrow trade unionists had started to look outward. Back in the 1950s, attempts to forge a national shop stewards' combine throughout the old Vickers company empire had met with the vaunted Barrovian insularity. Years later, a shop steward in the South Marston plant of the old Vickers empire recalled: "We tried, we tried a few times. But Barrow was always the stumbling block. They didn't seem to want to know about anybody else." (Beynon and Wainwright 1979, p. 79)

It was agreed at these meetings that the problems of defense industrial conversion had its sharpest focus at that time in Barrow, so it was decided that the Barrow Trades Council and TUCND would hold a national conversion conference in the town. The event, titled "Jobs at Risk" and held on September 22, 1983, attracted about 100 local, regional, and national union officials, members of Parliament, Labour defense spokesman Denzil Davies, and national and international conversion experts. "Jobs at Risk" was pivotal. It began with a discussion of Trident's perceived "first-strike" role and how the system differed from Polaris. Davies promised there would be no job losses if Labour canceled the Trident program. Urgent planning to replace Trident was called for. Gary Koos of the Electric Boat submarine yards at Groton, Connecticut (owned by the General Dynamics Corporation) described the economic dependence that was already being created by the huge demands of Trident submarine production in US shipyards.

Pronounced a huge success by the organizers, the conference was brought to a close by Danny Pearson. Speaking on behalf of the Barrow workforce, Pearson demanded that commercial uses be found for submarines and called for the workforce itself to devise new technologies. Referring to the newly formed Barrow Alternative Employment Committee, he said: "This meeting is just the beginning. If you have a good idea and technical data to back it up send it to us."[15]

Pearson later described the political difficulties faced by any group of people seeking even to raise the issue of alternative production in a community that had become so heavily dependent on defense contracts. As the senior TASS official and the chief union negotiator within the company, Pearson had recently taken over the job from Bill Latham, who he described in the aforementioned interview as "highly respected by both management and the men." But his predecessor, like many of the older trade-union officials in the yard, was not supportive of campaigning for alternatives; he felt that such an approach had been tried in the past and had failed: "He went on record regularly saying he wished he had a dock full of Tridents. . . . He didn't close his mind, but he did get up and debate for Trident within the branch and he'd quite a strong following with the more conservative members, but as far as the activists were concerned he was in the minority, the branch very much supported me and the part I played in the alternatives committee."

The conference launched Pearson and Terry McSorley (an official of "APEX," the Association of Professional, Executive and Computer Staff) onto what had become a "conversion circuit." They found themselves sharing platforms up and down the UK and beyond with former Lucas Aerospace shop stewards, politicians, and academics. As workers at a shipyard that had become virtually synonymous with Polaris and Trident, Pearson and McSorley were often invited to speak about the problems of plant-based conversion in general. They visited the US Trident shipyards at Groton (ironically once a part of the old Vickers global empire), and they took part in a conference on defense conversion in Boston. It was unprecedented for Barrow workers to join such a national and international campaign. Other members of the BAEC also traveled on this speaking circuit. Harry Siddall of TASS represented the BAEC at a conference on alternative industrial planning in Brussels at which a European network was set up to explore workers' alternative plans.[16]

Meanwhile, national trade unions, the Trade Union Congress, the Labour Party, and the Campaign for Nuclear Disarmament all began debating and adopting resolutions on conversion, and the subject gained more exposure. The higher profile of defense conversion was reflected in attempts by people on the "intellectual" and the "industrial" left to work together, Alan Milburn recalled when interviewed on January 26, 1993: "There was a tremendous spirit of optimism then, you were getting half a million people on marches, the world was going to change. But not only was there a lack of understanding in the trade-union and labor movement about conversion and CND, there was also a lack of understanding in CND about the trade-union and labor movement. Somehow we had to bring the two together."[17] The idea of raising funds to hire a researcher to investigate alternative production projects and markets emerged at a planning meeting in Barrow attended by shipyard shop stewards. Milburn recalled that it was decided that, if the BAEC was to have any chance of convincing the workforce that alternatives could sustain jobs, "we had to win the technical arguments." The Barrow Trades Council then sent Pearson to London to put the Barrow workers' dilemma to CND's national council. He recalled that he felt like his namesake, Daniel, going into the den of lions. Politically, he was being pulled in two different directions—putting forward alternatives to a potentially hostile audience at home in Barrow and then putting the case for protecting defense jobs to potentially hostile anti-nuclear campaigners gathered in London. This tension continued for the BAEC in what became a highly complex political arena, sometimes requiring great subtlety and humor. Pearson's impression of CND's council was indeed one of hostility, and he recalled only narrowly avoiding a row with one delegate. He told CND that for his members the choice had come down to "Trident or the dole," and he tried to explain that this wasn't any realistic kind of choice. He told them that if they cared so much about not building Trident, they might give the BAEC a researcher to help identify alternatives. "I told them," he recalled, "when the Bomb drops at least it's democratic, but unemployment is not like that, it falls first on the weakest members of society."

To the BAEC's surprise, CND offered a grant of £12,000 to fund a full-time alternatives researcher for Barrow. The grant was formally announced at the national anti-Trident demonstration organized by CND in Barrow later in 1984.

In January 1985, advertisements were placed in the *North West Evening Mail*, in the *Guardian*, in the *New Statesman*, in *Labour Weekly*, and in the *Morning Star* for a "Research Worker" to work with and for shipyard trade unionists to develop proposals for alternative production to Trident. This may have been the BAEC's most politically significant act. Some idea of its significance can be gained by contrasting the BAEC's ads with similar ads placed locally by the company around the same time. (See, e.g., figures 5.1 and 5.2.) Vickers, with its huge resources, was recruiting engineers in order to increase its skilled white-collar workforce to prepare for Trident. The BAEC, with its minuscule resources, was recruiting one person to attempt to turn those very new workers, thousands of their colleagues, the company, and the town in the opposite direction. In their stark divergence, it is apparent that the two advertisements represent networks that call upon vastly disproportionate sets of power relations and associations. We also know that the BAEC largely failed in its ambitious task. But stories about technology too often center around the powerful and the victorious. If instead we follow the "vanquished" in order to construct the technology's history, the narrative allows them a status nearer to that of the "victors." In this sense, it is easier to see technology as "know-how" embodied by people in their relationship with machines, rather than as something contained by artifacts. By not taking it as given that actors and networks differ in size, we can treat their inequality as a "product or effect" (Law 1994, p. 1).

VICKERS SHIPBUILDING & ENGINEERING LIMITED
(A MEMBER COMPANY OF BRITISH SHIPBUILDERS)

TECHNICAL VACANCIES

Vickers Shipbuilding and Engineering Limited, a major force in the Defence Industry, have a number of vacancies for qualified and experienced staff. The following disciplines are required:—

ELECTRICAL/ELECTRONIC DESIGN ENGINEERS
MARINE DESIGN ENGINEERS
MECHANICAL DESIGN ENGINEERS
NAVAL ARCHITECTS
SYSTEMS ANALYSISTS & ENGINEERS
COMMISSIONING ENGINEERS

Preference will be given to applicants with a degree and membership of the relevant professional body where appropriate.

Vickers Shipbuilding and Engineering Limited has an order book which provides security of employment on a long-term basis for the right applicants. The Company is due for privatisation within the next 12 months.

Please apply in writing, or Telephone for application forms to:—

Mr D. J. Stewart, Personnel Manager. Vickers Shipbuilding and Engineering Limited, PO Box 6. Barrow-in-Furness, Cumbria. Tel. 20351, ext. 5520.

Figure 5.1
Want ad in *North West Evening Mail*, early 1985.

There were 60 applicants for the research job. Five were interviewed. In April 1985, Steven Schofield, then a research student at Bradford University's School of Peace Studies, was appointed. Schofield had a B.A. in politics and modern history from Manchester University and an M.A. in defense technology and employment from Bradford. BAEC chairperson Terry McSorley said the resulting research report "must be widely read, so it must be in a form which is readable at the scientific and technical level and for the man or woman in the street."[18]

How the BAEC Represented Itself

From its outset, the BAEC agreed to stress economic and technological objections to the Trident contract, which was being negotiated as the BAEC was being constituted in 1984–85. According to Danny Pearson,

RESEARCH WORKER

To work with, and for,
Shipyard Trade Unionists in
Barrow-in-Furness,
to Develope proposals for alternative
production to the
Trident Nuclear Submarine.
12 month appointment.

Salary: £7,000.

Location: Barrow-in-Furness.

Application forms and further information
available from:
T. McSORLEY,
89 Marsh Street, Barrow;
and should be returned by February 25, 1985.

Figure 5.2
Want ad in *North West Evening Mail*, January 24, 1985.

the BAEC perceived the Campaign for Nuclear Disarmament as the pressure group that would "look after the moral arguments." Locally and nationally, CND was having considerable success in drawing crowds and winning ethical and political debates. A public meeting held in Barrow Civic Hall on July 25, 1984 was attended by 500 people.[19] But CND was to have no control over how the funds it donated to the BAEC would be spent or how the committee would work. It was partly to explain this independent position taken by the BAEC, and to account for how the money was being spent, that the committee so often sent its members on speaking trips to trade-union branches and peace groups all over the UK.

It was agreed that all members of the committee other than the research worker should be shipyard workers officially elected or delegated by their respective yards' trade-union branches. The BAEC was to be presented as a job-saving initiative. Its primary target audience was the workforce and the trade unions. Its other audiences were the local community, the Labour Party, the press, and the company's management. This was another potential pitfall. It soon became clear that the powerful local Confederation of Shipbuilding and Engineering Unions (known to workers as the "Confed") was hostile to the BAEC initiative.

The Confed saw its interests (and those of its members) as largely in line with winning the Trident orders. The first contract was being "sold" to the workers and the town as a "jam today and tomorrow" contract against the background of the national economic recession of the mid 1980s. The Confed was not represented officially on the BAEC committee, but its constituent unions—including the General, Municipal and Boilermakers Trade Union, the Technical, Administrative and Supervisory Section, the Association of Professional, Executive and Computer Staff, the Electrical, Electronic, Telecommunications and Plumbing Union, and the Amalgamated Union of Engineering Workers—were prepared to send delegates, thus avoiding open conflict with the Confed, whose secretary Frank Ward was known to be scathing about alternatives and told *The Observer* in an interview: "Some of them want us to make electric rocking chairs. Or convert to kidney machines. If that yard went into proper production of kidney machines, there would be enough for every bugger in the country inside a fortnight."[20]

Both the manual workers' and the staff employees' shop stewards' committees in the yard agreed to send representatives. Ironically, some of the BAEC's most active members were also Confed officials. Danny Pearson was TASS's senior negotiator within the yard, and Harry Siddall was an APEX convenor. Together they represented increasingly influential sections of the workforce. (Many members of TASS were designers, planners, and draftsman, and technical and white-collar staff employees began to outnumber manual workers as a result of the increasingly sophisticated nature of nuclear submarine technology). In steering this careful course around the Confed and around more conservative trade-union members, the BAEC agreed to have no formal links with CND locally—that would have been the political "kiss of death," according to Pearson.[21] Even nationally, the relationship was kept at arm's length, contact being maintained mostly through Alan Milburn at TUCND, who acted as a bridge between unilateralist "hard liners" who would cancel projects first and talk about jobs second and moderates in the industry who would put jobs first and disarmament second.

Another obstacle was the Barrow shipyard's management. It would have nothing to do with the BAEC. Invitations sent to management to attend BAEC meetings were refused. Schofield, the researcher, found that he could not get management to cooperate with him on his first task: compiling an up-to-date profile of the shipyard's and the engineering works' skills and resources.

Constituting the BAEC was therefore a delicate political process involving balancing many strands of opinion among workers, community, management, and political and pressure groups. In an interview conducted on October 20, 1992, Schofield explained this problem as follows:

There were people that were in the Confed like Danny . . . and he was well aware of the problem because he had a lot of respect for certain individuals on the Confed, but they were fundamentally opposed to the argument we were putting forward . . . and he had to try to balance his membership of the Confed with being on the committee and that was why this argument about not being opposed to defense work became so central for the committee . . . to try and offset this argument that we were basically peaceniks who didn't have any credibility. . . . There had been peace movement people who'd come into Barrow and given presentations to the Confed . . . the sort of legacy I had was people who'd come in saying well er . . . we want you to build peaceful products not war products and they'd said well, what do you suggest? . . . And they said things like washing machines—they were *seriously saying* to people who worked at Vickers they should be building washing machines instead of nuclear submarines. . . . Now that obviously was totally inappropriate.

The BAEC decided that it would use the economic and technical arguments in its attempt to steer the shipyard away from dependence on monolithic MoD orders. A more varied order book based on developing existing engineering skills and shipyard resources would be presented as the key to a more secure future for the employees. And the huge financial cost of Trident would be represented as a gamble with the town's future. The Trident contracts were already fraught with uncertainty for the community. It was widely believed in Barrow at the time that the election of a Conservative MP in 1983 had been a political "hiccough" and that the traditionally Labour-voting town would revert to Labour in the June 1987 general election. If this switch to Labour were to be reflected nationally, the Trident contract might be halted, or even canceled altogether, in line with Labour's policy of non-nuclear defense. The yard might be left high and dry, with no other orders. But if Labour failed to form the next government, so the argument ran, the Trident program, with its escalating costs, might be such a drain on the nation's defense budget (in a period of recession) that it would be vulnerable to early curtailment by a Conservative government. "Neither the West nor the Eastern bloc could sustain this type of expenditure," Terry McSorley argued when interviewed on July 31, 1992. That this colossal economic risk was bound up with the Trident contracts was reflected in the reports of wrangles over cancellation-compensation clauses in the contracts that were surfacing during the privatization of Vickers Barrow. The yard's true "value" became extremely hard to assess because of the political uncertainty surrounding the "core" business.

Technological Conservatism

Around 1985–86, evidence was brought to the BAEC that Barrow's engineering capacity was being shelved as the company began to throw all its resources and its entire workforce into Trident submarine production. The portfolio prepared by Lazard Brothers for the company privatization stated that in 1985 Vickers held more than 30 full or provisional patents in the UK and overseas. But it appeared that few of the Vickers-held patents had been exploited. In addition, there were eight existing license agreements for the manufacture of Vickers-designed products.

The BAEC used what it saw as the trend toward technological conservatism and the perceived risks and uncertainties associated with Trident to question management's rhetoric about Trident's capacity to secure jobs for the whole workforce into the twenty-first century. At the same time, VSEL's chief executive officer, Rodney Leach, used the launch of the submarine *Trenchant* to appeal for workers' commitment and loyalty to the Trident network. Leach urged Vickers workers to "have faith and confidence" in the Trident program. He condemned suggestions that anything other than Trident would fully occupy the Barrow yards workforce for the next decade and more as "mischievous" and "misleading." "No one," he said, "should be in any doubt that the single best prospect for improving company performance while maintaining security of employment in this works and in this town lies in completing the Trident program. I ask all within the company to have faith and confidence in the Trident program. If we ourselves fail to support Trident how could we expect others to do so?"[22]

It could be argued that the Barrow Alternative Employment Committee was launched at a time that was economically favorable for its argument that a full shipyard order book could be used as a springboard for developing alternative markets to sustain the industry when orders ran out. Politically, however, the BAEC was launched at the worst possible time, insofar as the Trident program appeared to secure employment. The local political climate for alternatives became more favorable in the early 1990s, when the promised job security ran out far earlier than the company's "sales" rhetoric had predicted, making Trident's negative effect on employment more apparent. Many of the BAEC's warnings of the mid 1980s were realized in the early 1990s, but by then the economic climate was less favorable to fostering technological change.

Conclusion

Workers' plans for alternative production are almost always judged in "political" terms, as responses to the BAEC's plan later showed. Managerial authority is fundamentally challenged by workers' plans, which (somewhat ironically) also challenge the traditional concerns of the trade-union movement. Just as the Lucas Aerospace Combine discovered when campaigning for its alternative plans, the BAEC found itself swimming against the tide of both management and old-style trade-union orthodoxy (Wainwright and Elliott 1982). The BAEC's proposals were therefore not judged on their "technical" merits; they were dismissed out of hand as "dangerously misleading to the point of being mischievous."[23] Though the BAEC wanted to "win the technical arguments," it was never taken seriously as a technological actor. Here the decision as to what is technical and what is social is crucial. The BAEC was unable to win the technical arguments because it did not build a sufficiently technological identity with which to enter the technopolitical debate.

In order for the Trident network to stabilize, the "official" Trident builders (i.e., the company) had to be seen as the real technological actors. It followed, then, that the BAEC could be cast as misleading and mischievous because it had a "political" agenda.[24] Once it was cast in that light, the specific technologies it advocated did not have to be examined in detail, and their technical feasibility never became an issue. Serious debate of the BAEC's recommendations would have required removing "technical feasibility" from its occluded "apolitical" domain (a domain occupied by management) and allowing shipyard products to be discussed openly, with actors such as trade-union convenors expressing their views. It would then have been clear that the "technical" was also "political." The company would not allow this to happen, even though, as Terry McSorley pointed out in his interview, "it was the very *technicians* which the company employed who were producing the arguments for alternatives," and "Rodney Leach was the *politician* presenting the company case."

6
Building a Counter-Network

The Barrow Alternative Employment Committee was formed in September 1984.[1] Before its official launch, several of its members had been invited to take part in a United Nations-sponsored conference on arms conversion in Boston. Former Lucas Aerospace shop steward Phil Asquith had been asked to form a delegation that could speak from experience on problems of diversification in the UK, and he had asked the Barrow Trades Council for representatives. Danny Pearson and Terry McSorley joined the delegation, along with the general secretaries of the main UK and European industrial trade unions, academics, and representatives of organizations active in the UK peace movement, such as the Trade Union and Christian Campaigns for Nuclear Disarmament. Ron Todd, the leader of the Transport and General Workers Union, which represented large numbers of defense workers, attended the conference because the TGWU was developing a policy with which to respond to the large defense cuts that were expected in the UK. Although the TGWU had very few members in the Barrow shipyard, it was the TGWU that would later give the BAEC the funds that enabled it to publish its final study, Oceans of Work.

Apart from local Boston defense workers, McSorley and Pearson recalled that they seemed to be the only "grassroots" contingent at the conference, and consequently were in demand as speakers. They visited trade-union groups in Groton, Connecticut, where the US Trident program was in full swing. They found that the political climate at Groton was polarized in a way that never quite occurred in Barrow. Terry McSorley, interviewed on December 2, 1993, recalled: "What they couldn't understand was why . . . or how we'd managed to get cooperation between peace groups, CND and trade unionists in the yard. The debate in the UK seemed to be more politically sophisticated, they had no idea that there could be cooperation between so-called peace groups and the workforce which could be to their mutual benefit . . . these were two sides of an argument which would never meet."

The trip to the United States gave the Barrow movement an international information resource. Valuable contacts were established, not only within the UK but also with organizations such as the US Office of Economic Adjustment. These contacts later gave the BAEC confidence in pursuing difficult arguments at home—especially within the Labour Party, which was developing a new defense policy. The BAEC and other trade-union alliances later influenced the Labour opposition to agree to creating a defense diversification agency to respond to shrinkage in the defense industry once Labour returned to power.

How the BAEC Was Constituted

The first of the minuted meetings dealt with the election of officers and the constitution of the group. Present were Terry McSorley, a Vickers timekeeper and an APEX staff representative; Bob Bolton, a pipe welder and an official of the GMBTU; Dick Wade, an electrician, representing the EETPU; Harry Siddall, a chief technical clerk and an APEX staff representative; Les Gallagher, an electrician, a member of the Barrow Trades Council, and a former EETPU activist (he had resigned in protest of his union's strike-breaking activities at Wapping); Danny Pearson, a project planner and TASS convenor; Roy Silcocks, also representing TASS; and the committee's new researcher, Steven Schofield.[2]

The minutes of that first meeting revealed fundamental problems that the group would encounter over the next 2 years. Though financial matters took up much of the formal discussion throughout the BAEC's existence, the political dilemma of gaining official acceptance within the shipyard and the community would prove to be the most frustrating aspect of the campaign.

Throughout the period in which the BAEC was active, there were dual demands on it and on Steven Schofield. These resulted from the group's association with emerging local and global networks. The campaign had to build a credible local presence and identity, yet also maintain a growing national and international profile. For example, the first meeting heard that Schofield had just appeared on a program about defense industrial conversion on Danish television yet had also been interviewed by Radio Furness, the Barrow-based BBC station, about his appointment and his task. These two interviews addressed divergent audiences. Although the BAEC was a locally based initiative, it was the local audience which was harder to "win" and with whom at times it proved the most difficult to engage, notwithstanding the contrast with labor politics at US shipyards. The BAEC 's basic message—that Trident would actually destroy jobs even if it created employment in the short term—was the hardest message to communicate. The BAEC's analysis that Trident would have a negative local economic effect had been underlined by Denis Healey, then the shadow foreign secretary, speaking in Barrow at a public meeting organized by the APEX union: "The multi-billion pound Trident program could rob Barrow of its bread and butter work. . . . The best way to preserve the long-term job security of yard jobs was to continue with existing traditional programs."[3] However, the task of making the arguments locally, on a regular basis, required a different kind of political skill. Healey didn't have to live and work in Barrow.

There were many discussions about how to set up a meeting with the shipyard trade unions' umbrella organization, the Confederation of Shipbuilding and Engineering Unions. Schofield had written officially requesting this, but had received no reply, so it was agreed that Pearson (a TASS convenor and also a Confed committee member) would personally deliver another letter at the next Confed meeting. It was important to tread carefully. Meanwhile, the individual trade-union branches were proving more amenable. Meetings were set up at which the BAEC could speak, outline its case, and appeal for support. And other avenues would be tried. It was agreed to contact the Workers' Educational Association about the possibility of input into local trade unions' educational courses. Welfare State, a local arts group, was approached to discuss fund-raising initiatives and other ways of raising the local profile.

These were all routes through which workers, their families, and the local community might be reached. Some of the methods of communication that the BAEC tried were direct and used official channels, such as the approach to the Confed; others, such as theatre groups, were more oblique. At the inaugural meeting, the most direct and most fruitless approach of all was discussed: sending a letter to VSEL's chief executive officer, Rodney Leach, to "ask permission to visit the yard to look at various areas of production."[4] By the time of the next monthly meeting, the BAEC learned that this request had been turned down. The reply, signed by Leach, basically said that management was monitoring the position regarding new products and did not see the need for any "outside" interference. Steven Schofield, interviewed on July 29, 1994, said: "I think they saw us as an irrelevance. But we had to show a willingness to cooperate and not to threaten the management." The management response was quoted in a feature article introducing the BAEC in the *North West Evening Mail*. Management had said that it was not appropriate for the company to be involved in the BAEC project "given the special sensitivity of VSEL as a public sector concern actively engaged on major British defense service contracts."[5]

That appears to have been the first and last direct, formal contact between the BAEC and VSEL. The BAEC agreed that it would take no further steps to involve VSEL management until it had obtained its crucial meeting with the other reluctant "official" actor in the shipyard organization: the Confed. Once the BAEC had received the hoped-for response from the Confed, Schofield would attempt arrange a meeting of all the shipyard's shop stewards. It was hoped that this would strengthen the campaign's political position within the yard and its standing with management. It was decided that the timing of all these approaches was a delicate matter and that existing protocol should be followed.

Meanwhile, the Workers' Educational Association had responded that, though it had no objection to the BAEC's work, it would not be permitted to address WEA classes, out of the fear that to do so would lay the WEA open to "dangerous" approaches from "other bodies."[6] Welfare State replied that, depending as it did on government grants for its survival, it could not be seen to be supporting the BAEC, but it could offer to provide entertainment for fund-raising events. It seems that the BAEC soon became cast as a "political" actor rather than a "technical" or "educational" one. Thus, while tacitly supporting its aims, local organizations found the BAEC difficult to accommodate explicitly.

The BAEC's only assured income was the one-time major grant from the national CND. Accordingly, it was decided that a national promotional and fund-raising leaflet should be produced that would appeal to churches, trades councils, local disarmament and peace groups, and voluntary bodies for donations. Separate, appropriately worded letters trawling for donations would go to all branches of trade unions and to the unions' district and county councils. However, the main business of this August 1985 meeting was how to audit the skills and resources of the shipyard despite the official rejection of assistance and access by VSEL management. Without detailed information on skills and resources, the researcher could not begin to identify other products and markets that would have any local relevance. Each BAEC delegate therefore agreed to compile a report on the numbers of workers, trades, types, and functions of machinery in his own work area. The data could then be collated to assist the research. A newsletter about the aims of the BAEC was to be produced over the next month for distribution among the workforce. It was hoped that this newsletter would encourage informal responses and increase the flow of information from workers to the BAEC.

But the shipyard was in the process of privatization, and for the trade unions this was a time of huge uncertainty. The yard was to be sold off before the next general election, which was to take place in June 1987. The privatization and the outcome of the election were parts of the political platform upon which the Trident contracts rested. The BAEC meanwhile decided to carry out its own analysis of the privatization and to make the results of that analysis the subject of its first official publication. No copy of the document seems to have survived, but members recalled that it was intended to provide alternative, non-managerialist information to the yard trade unions, including business profiles of the merchant bank handling the sale (Lazard Brothers) and the possible bidders for the yard (GEC and Trafalgar House), plus a discussion of the implications for employment and production. "There was very little access for the trade unions to that sort of material," Schofield recalled. "So we made it our first project."

Late in 1985, Schofield was given a slot at one of the Confed's regular meetings. "I was listened to politely," he recalled. "I gave a straightforward industrial analysis of conversion. But Frank Ward, the secretary tried to turn it into a political thing, you know.... 'You're CND, you lot just want us to build washing machines.' The Confed had held a meeting years before with some Quakers who unfortunately had suggested just that. Other contributions from the floor were interesting but basically people could just see no prospect of any change, though some people did recognize there was a problem." After this, the BAEC recognized that there would be no official support from the Confed.

Donations from local union branches had begun to arrive by the next month, along with an invitation for the researcher to take part in a debate about Trident and the privatization of Vickers organized by the Barrow and District Engineers Association. The Barrow Borough Council requested a financial report from the BAEC to help it decide on the BAEC's application for a grant. The *Evening Mail* reported: "Labour councillor Maurice Kerr said he felt 'professionals' could come up with the answer [on alternatives] much more quickly, and the council should not spend money on 'an amateur.' He was backed by Conservative leader Ted Smith."[7] This questioning of Steven Schofield's credentials drew a quick response from Terry McSorley, who said the councillors were ignoring the "expert backing" Schofield brought from Bradford University:

"You wouldn't get that sort of backup for the money that is available any other way. It is a tried and tested method of producing research and councils themselves have used it in the past."[8]

With input from BAEC delegates, a skills profile of the yard began to come together. It is interesting to consider how such a profile, compiled unofficially, would have differed in content from the official account of VSEL's activities contained in the company's glossy literature and sales brochures. For example, the BAEC's minutes of September 3, 1985 refer to copies of minutes of VSEL's former internal Monitoring Committee that the BAEC had obtained. Abolished on privatization, the Monitoring Committee was a management-worker forum that had been started up by British Shipbuilders (under nationalization) to discuss production problems and possible solutions. The BAEC realized that these minutes would be useful for helping to identify engineering projects that the company could have been pursuing—particularly commercial products such as the Constant Speed Generator Drive.

That existing skills and resources should be directed into new markets and technologies was the policy line to be stressed whenever BAEC members gave interviews to newspapers, radio, or television. And when it came to talking about Trident, the cost in employment and economic terms was to be emphasized. The BAEC's stance should reflect its own unique position and must be both practical and credible. It would be easier to convince the workforce and community about the value of alternatives by emphasizing costs and employment risks than by emphasizing "the immoral values of building the (Trident) system."[9]

In July 1985, the Bradford School of Peace Studies published the results of a study of employment in the UK's defense industry by Peter Southwood. Southwood warned that there would be large-scale job losses when workers were no longer required on MoD programs. He also predicted that the trade unions would find themselves forced to lobby for more defense orders in the absence of industrial conversion programs. Workers would find themselves either sacked or, ironically, pressing for continued dependence on defense work. Workers at Vickers Barrow were in the most vulnerable employment position in the UK (Southwood 1985). The BAEC took the opportunity to respond to this report, thereby provoking a counter-response from VSEL. Terry McSorley told the *Evening Mail* that the BAEC "had to get its program right in order to carry Vickers's unions with it," and that "alternatives had to be ready for implementation if Trident was canceled."[10] But a Vickers spokesman responded that defense contracts had played the major role in keeping the workforce in regular employment for 25 years. Traditional skills would not be lost by concentration on Trident, it was claimed. This rare statement from Vickers management contained what could itself be seen as a description of baroque (complex yet conservative) technology: "In many respects the Trident vessels are simply larger versions of the boats we currently build, employing many of the same techniques in their construction."[11]

The BAEC was aiming to produce two major reports. The first would lay out the financial and employment costs of Trident; the second would be a "technical" study reporting the results of the BAEC's research into employment-saving alternatives.

Meanwhile, there were numerous meetings to attend as the network became wider and more complex. Schofield and the BAEC delegates were simultaneously lobbying the Labour Group on the Barrow Borough Council for local political and financial support, sending speakers to national CND conferences, meeting environmental researchers to talk about wind power generation, being interviewed by Swedish TV and newspapers, meeting Martin O'Neill (Labour's defense spokesman), and attending crucial meetings with the Barrow shipyard's numerous trade-union branches. Contacts were being developed with former Lucas Aerospace workers, and a meeting was arranged with Vickers workers at Elswick on Tyneside, who were considering reviving their own conversion efforts.[12] (Vickers was now a separate company.)

At the December 1985 meeting, it was reported that 3000 copies of the BAEC's first newsletter had been distributed around the shipyard. It was agreed that officials ask for comments on this newsletter at each union branch meeting, and that a second newsletter would be issued in January. Workers would have to be kept aware that the BAEC was still functioning as a workers' committee while the research was being carried out. Ways of canvassing yard workers for their opinions of the Trident contracts were discussed. (As far as anyone knew, this had not been tried directly before, but a previous study based on a street survey of local attitudes toward defense work had shown ambivalence toward the program in the community.) The street survey was conducted by the Barrow CND group. The results (later analyzed at Lancaster University's Richardson Institute of Peace and Conflict Studies) gave some indication that local people were, in common with many people at the time, against new nuclear weapons systems such as Cruise and Pershing, yet either in favor of or ambivalent about Trident. While this obviously related to uneasiness about Barrow's role as "Trident Town," it may also have had something to do with political and media representations of Trident as the technological and political successor to Polaris. In spite of its US origins, Trident could be presented as the UK's "independent nuclear deterrent," especially in the town that was building the UK-designed platform (i.e., the submarine). Cruise was more obviously a US system based in the UK. The BAEC, in its interim research report, tried to undermine this distinction by drawing links between Trident and the Cruise and Pershing systems, presenting them all as elements of an increasingly dangerous and unpopular move by the United States toward the concept of "winnable" nuclear war staged in the European theater. The latter view of Trident, it was thought, would challenge the loyalty to the project of Barrow workers, many of whom cited national security as the overriding reason for "doing the job" (Smoker 1985).

Meanwhile, the national CND produced a fund-raising leaflet intended solely to benefit the BAEC. In this leaflet, the CND, then at the height of its national popularity, played down its own role in the Barrow initiative, recognizing the special political difficulties that Barrow workers pushing for alternatives faced. Funds to sustain the shipyard campaign were getting tighter, but donations from diverse sources continued to arrive and continued to be raised as members spoke at meetings and conferences such as the one on "A Better Future for Defence Jobs" that Harry Siddall addressed in Southampton. (A technical clerk who worked on conventional submarine programs, Siddall became one of the BAEC's most active speakers. The minutes for February 1986 contain his report on addressing the Derbyshire Miners' Committee about the campaign. April took him to Brussels to speak at a shipbuilding conversion conference.)

Schofield produced the privatization report. One of his many letters published in the *Evening Mail* warned that whoever bought the company would own the most defense-dependent firm in the country, and that crucial chances to diversify and to set up alternative markets were being wasted by the selloff efforts.[13]

The difficulties inherent in building a local identity while maintaining a growing national profile were becoming evident. Activities mentioned in the February 1986 minutes include attending the annual meeting of the TUCND, participating in the Labour Party's Defence Study Group in the House of Commons, speaking at a public meeting of the Peaceful Production Research Political Unit in Durham, attending a conference about Trident in Birmingham and a youth trade-union conference in Manchester, and organizing a "day school" on Labour's defense policy in Barrow. Danny Pearson, interviewed on November 6, 1991, commented:

"It was always a trap that you could fall into, getting carried away with your own notoriety like." And while all these contacts were being made and these alliances being formed, the central problem—gaining acceptance among all groups inside the shipyard—remained obdurate. With management and the Confed either hostile or indifferent to the initiative, maintaining local credibility became a serious problem for the BAEC. Pearson explained: "It made it very very difficult for the committee to retain the membership . . . not only the membership of the committee, but the feedback into the branches. . . . It was only because the people we had on the committee had the respect of their branches—the branches knew from their pedigree what kind of people they were and that tended to stand them in good stead through the rough period that the committee had."

Local and national elements of the campaign were brought together when Martin O'Neill, Labour's shadow defense spokesman, visited Barrow. O'Neill met with Vickers management, with trade unions, with the BAEC, with local Labour officials, and with Barrow CND. Each of these groups had its own identity, but they had many members in common, owing to the closeness and interdependence of the town and the shipyard. O'Neill was trying to build support for Labour's non-nuclear defense policy. If Labour could retake the Barrow and Furness constituency in the forthcoming general election, it would probably be forming the next government. It was essential for Labour somehow to persuade people that canceling Trident would not have to mean wholesale job cuts. Alternative products and future conventional weapons orders were discussed; however, for Labour the answer seemed to lie in renouncing the Trident missiles yet still building the submarines. O'Neill was quoted as saying that he had "discussed with management the possibility of using Trident submarines purely as hunter killer submarines without deploying the nuclear warheads."[14] Whatever management's private views of this radically altered role for Trident were, publicly the idea was endorsed as technically feasible. Endorsement by VSEL was expedient largely because the yard's privatization was underway and one of the greatest "risks" affecting prospective buyers probably was that a new Labour government would cancel the Trident submarine program the following year. The "risks" of cancellation therefore had to be officially played down.

The BAEC's March meeting was attended by Schofield's two research supervisors from Bradford University, who suggested that the committee's first major report could be produced as one of the School of Peace Studies' Peace Research Reports, which already had an established profile and a distribution network. They believed, however, that the final report should be issued separately by the BAEC. The following passage in the BAEC minutes of March 11, 1986 reflects fears about the effect on local credibility: "At this point the Committee agreed to the suggestion of letting Bradford Univ. issue the publication of the report, so long as it still had the BAEC displayed on the front cover. A cover price of between £1.25 and £1.50 was set for the report and press releases to the national newspapers were left in the University's hand." This decision was practical and convenient, but it may have been unfortunate and at odds with the BAEC's original aim of maintaining its local workerist identity. While running its activities on a shoestring, the BAEC had to weigh the advantages of accepting support from Bradford University against the possible strategic disadvantages of loss of local control and local credibility. The first report (entitled Employment and Security—Alternatives to Trident) was widely dismissed in Barrow as "politically" motivated and as influenced by "outsiders." More important, because of what in hindsight looks like poor timing, the interim report appears to have overshadowed the final, "technical" report (Oceans of Work).

The March minutes recorded more donations, appearances at meetings and conferences all over the country, and discussion of the persistent problem of obtaining a comprehensive skills audit of VSEL. Apart from access problems, this was complicated by the changing nature of the yard's facilities and workforce in the run-up to Trident—notably the construction of the huge £300 million Devonshire Dock Hall, the covered submarine production shed, and the recruitment of record numbers of white-collar workers. The BAEC members agreed that the report might have to contain a simple overview of skills and their possible new applications.

A proposal by local anti-nuclear activists to set up a "peace camp" near the shipyard caused some alarm. This was precisely what the BAEC feared most, believing that it would undo all the painstaking explaining that had been done. In the event, the peace camp never took place. The BAEC minutes of March 11, 1986 contain this rather crisp entry: "T. McSorley reported on a telephone conversation he received in relation to the establishment of a peace camp, upon which he poured cold water."

Jobs at Risk

In 1986 the BAEC published an interim research report, described as a "non-technical" document aimed at outlining preliminary findings and creating the context for a detailed final report on alternative-employment-sustaining technologies (Schofield and BAEC 1986). This report (titled Employment and Security—Alternatives to Trident) began by acknowledging an important limitation: "The refusal of the management of Vickers to cooperate with the BAEC has hindered the development of a comprehensive listing of facilities and workforce numbers, as well as up to date information on the company's products and services." The report reflected the widely held expectation of an impending change of government and hence a non-nuclear defense policy operating from the political center. It aimed to "stimulate debate, not only about Barrow but in the wider context." "By the time of the publication of our final report," the authors announced, "we expect work on conversion to be underway on a comprehensive, national basis in which BAEC can continue to make a major contribution."

The interim report emphasized the cost of the Trident system. The official cost of the system, initially set at £5 billion, had rapidly risen to £7.5 billion when the UK had gone along with the US decision to go for the enhanced (i.e., more accuracy, counterforce capable) D5 system—which required larger submarines and larger construction and maintenance facilities—rather than the C4. By January 1985 the cost had risen to £9.285 billion (Schofield and Barrow Alternative Employment Committee 1986). Yet, as cost rose, projected employment on the system in the UK actually fell. Initial attractive estimates of 25,000 jobs and another 20,000 in support industries turned into 9000 direct and 7000 indirect jobs, said the report. Because of this, the overall cost of creating each job (including research and development costs) on Trident was put at between £309,000 and £580,000 at the project's peak. Civil jobs, the BAEC contended, would be far cheaper to create.

The report then examined the consequences for the shipyard of a non-nuclear defense policy that would lead to the cancellation of Trident, in line with current Labour policy. Instead of fudging the local consequences of Labour's policy, the BAEC believed, it was necessary to confront and clarify what was proving to be the party's most controversial pledge. Rather than represent Labour's policy as the risk (a tactic of the pro-Trident lobby), it cast the Trident program itself as the risk. It underlined the economic and technological uncertainties contained in adopting the Trident program, but it also emphasized the political risk of undertaking the Trident contracts, particularly in the run-up to the 1987 general election.

Here was another political dilemma for workers supporting alternatives: the BAEC clearly supported Labour's non-nuclear defense policy, yet it knew that the policy could easily be presented as the major threat to jobs in Barrow. So the BAEC turned the argument around. Rather than take Trident as the given, it treated the submarine program as contingent—as an unstable project. That, then, was the risk that had to be borne by workers, by the community, by the country. However, by supporting adoption of a non-nuclear defense policy, and therefore supporting cancellation of Trident, the BAEC's interim report itself risked moving away from the BAEC's strategy of emphasizing the economic and technological arguments against adoption of Trident (i.e., that it would put jobs at risk) and engaging with the ongoing fierce party political debate. In detailing the moves toward "first-strike" capability and the concepts of "battlefield" and "winnable" nuclear war that were being adopted for Trident and also for Cruise and Pershing by US and UK governments, it can be seen that to talk about Trident in this way was to engage in global politics. The BAEC could hardly avoid such engagement in an interim report about such an overtly political technology. In this sense, while the production arena was still defense technology, the political climate was arguably "hotter" for the BAEC than it was for the Lucas Aerospace shop stewards' combine.

The interim report also argued that the demand for conventional submarines, surface ships, and non-nuclear weapons would increase as a result of a transition from offensive to defensive policies under Labour's proposals. Though the Barrow community could expect to receive some of this work, shipyard conversion and product diversification away from defense would still be essential for long-term job security there. Loss of employment resulting from the adoption of a non-nuclear defense policy should not be compensated for by a wholesale increase in spending on conventional defense equipment, in the BAEC's view. Defense workers also needed a program of alternative, civil products to reduce their overdependence on one customer. According to the interim report, "the most effective policy in these circumstances would be to convince defense workers that their skills and abilities will be more effectively used in non-defense work and that the government is committed to shifting considerable resources, in terms of investment and research and development, to new civilian programs" (BAEC 1986, p. 38).

The capital-intensive and labor-intensive nature of projects such as Trident created a threat to long-term employment that could only then be preserved through continued massive levels of investment—levels that were not likely to be sustained. A choice about future employment had to be made. Either the workforce should accept the present reliance on defense work as a means of sustaining employment or Vickers should begin to identify major new areas of civil R&D and production (for example, in maritime technology) and orient its production and employment patterns toward these areas while not totally abandoning its base in naval defense contracting. Relying on exports as a means of sustaining jobs might be even more precarious than relying on MoD orders.

Defense Conversion, the Lucas Plan, and Technopolitics

The BAEC's interim report went on to describe conversion activity in the United States, then to outline the corporate plan of the Lucas Aerospace shop stewards (which it described as the "best example yet of an industry based conversion program." The Lucas Plan, launched in 1976, was undoubtedly an important influence on the BAEC. There were many parallels. Both initiatives sprang from dynamic unofficial or semi-official trade-union campaigns. Both campaigns were extremely popular nationally and internationally yet often struggled to keep their workforces and their official trade unions on board. By engaging with the *content* of production, both transcended accepted boundaries of trade-union activity and acted to promote blue-collar–white-collar cooperation. Both were spurned by their managements, politely overlooked by the Labour Party, and largely ignored by the respective national trade unions. And, in particular, both were victims of Confed inertia.

But there were also differences. Industrially, geographically, and in terms of trade-union activity, Vickers/VSEL was substantially different from Lucas. VSEL was much more closely tied to the Ministry of Defence and the Royal Navy. And the BAEC was wary of getting embroiled in what it saw as the labor politics of alternative plans. The interim report says: "The emphasis we place is on the industrial element of the [Lucas] Plan as opposed to the history of the Combine and the political and trade union struggles to have the Plan supported and implemented."

Perhaps, in concentrating on trying to win the "technical arguments," the BAEC neglected the political engineering that was necessary if its arguments were to be heard. There is, however, little evidence for this, because to talk about alternatives—indeed, to question the Trident program at all—was always highly controversial. As Terry McSorley explained in an interview on July 31, 1992, the BAEC was trying to avert redundancies that it knew, because of the very nature of the Trident project, were bound to emerge, but which hadn't yet become evident; the Lucas Plan, in contrast, was an initiative to stem the tide of redundancies that had already begun in the Lucas empire.

It could be argued that the national political and economic climate had been more favorable for the Lucas workers operating under the last Labour government of the 1970s and taking place in an era of public ownership. Because of the economic cushioning effect of MoD contracts, Vickers/VSEL had been partially insulated from the industrial recession that later hit the UK. It was easier in Barrow to hold the view that mass redundancies would not happen than it would have been to maintain such a position elsewhere. Significantly, during the life span of the BAEC there was full employment in the yard, and the local economy appeared robust. People had a lot to lose. The political and economic climate had changed since the Lucas initiative. Ideas about industrial democracy, popular in the 1970s, had given way to recessions, mass unemployment, and falling trade-union membership. Presenting a defense specialist company with a wide range of potential new products in a recession was unlikely to work, said the interim report. In identifying new products, it was necessary to consider the current national manufacturing decline and the practical possibilities for conversion in a recession. It was necessary to recognize from the outset that the barriers to conversion were "not technical, but political and attitudinal, requiring government and industry support and planning" (Schofield and BAEC 1986, p. 58).

There is an ambiguity here. The BAEC was keen to avoid being seen as an overtly political or even revolutionary movement, yet the report acknowledged that the main barriers to a return to general engineering products or to the adoption of alternative technologies were political and social. According to Wainwright and Elliot (1982, p. 51):

Many reasons were put forward by [Lucas] management for its subsequent rejection of all product ideas, some on the basis of profitability, others because of technological considerations. Above all else though, was the Company's strategy of specializing in defense work for the aerospace industry, although management claimed that prospects for diversification were constantly reviewed. . . . In order to maintain its specialism in defense work the management's corporate strategy required rejection of the Plan. Power relationships would seem to have been more important than the possible future prosperity of the company.

This passage could almost have been written 10 years later about VSEL's subsequent rejection of the BAEC's alternative production proposals. Among the similarities in the two stories are the companies' strategy of maintaining defense specialization (later embodied in the VSEL catch phrase "core business") as the rationale for production. On one hand the alternatives campaign wanted to concentrate on technical possibilities; on the other hand it recognized that a wider context was necessary for technical ideas to be accepted. The BAEC was, in effect, entering the negotiations about what is seen as social and what is seen as technical, and about who gets to decide how and where a line is drawn.

The BAEC's interim report listed the civil engineering products and vessels that had been built at Barrow in the past and the diverse skills that these had generated in an unique workforce. Management was criticized for allowing a culture of dependence on the Ministry of Defence to develop and for its inability to take risks in developing products already being pioneered in house. Commitment to defense requirements had promoted caution about the "uncertainties" of the commercial market and conservatism (even negativism) about the potential for new product development. Here the report mentioned two Vickers inventions—the Constant Speed Generator Drive and the Fuel Homogeniser—as examples of negativism toward innovation. Success in the civil sector depended on a positive management outlook wedded not to the bureaucratic certainties and practices of Ministry of Defence production but to the dynamic elements of commercial success. The Barrow yard, the largest integrated shipbuilding and engineering site in Europe, had the skills and the facilities to develop a successful civil marine industry. This, the report maintained, could include work on renewable energy—ocean thermal energy, tidal power, wind and wave power. Also highlighted in the report were undersea mineral extraction and civil submersibles for offshore work and exploration. So the report questioned the Trident contract and set out an alternative role for the shipyard that drew on its recent and existing engineering expertise, projecting these into new marine-centered markets.

After some delays, the interim report was published late in July 1986, after the privatization of the shipyard, the management buyout, and the workers' share offer. May and June had seen 6000 manual workers on strike over pay and conditions and 5000 white-collar staff employees refusing overtime and working to rule. BAEC members had become involved in these disputes in their individual trade-union roles. These immediate problems and distractions had often pushed campaigning for alternative production within the yard into the background.

The interim report was distributed by the Bradford School of Peace Studies, but the BAEC decided to do its own publicity and to call a press conference of all the local newspapers and other media. Every trade union that had donated to the project would be sent a copy. To keep the momentum going, it was decided to hold a follow-up conference to discuss the issues raised in the report.

Responses to the Interim Report

The publication of Employment and Security—Alternatives to Trident was given front-page coverage in the *Evening Mail*. By a strange coincidence, however, it coincided with the Ministry of Defence's official invitation to VSEL to bid for a second Trident contract. This undercut the BAEC's central message: that Trident was a threat to jobs.

The story on the BAEC report ran as follows:

> Winning power from the sea with a Morecambe Bay barrage and building submarine oil tankers are among a new report's suggestions for switching Vickers away from Trident.
>
> The 100-page report was published on the day VSEL was invited to tender for the second Trident contract. . . . The report follows 12 months of investigation by Barrow Alternative Employment Committee researcher Steve Schofield, who hopes it will be read not only in Barrow but throughout the country. . . . But even before seeing the report Vickers said: "We have no evidence there would be any benefit to the company if we were to embark upon some of the schemes which Mr Schofield suggests."[15]

The *Mail* then devoted an inside page to the detailed "findings" of the report, acknowledging the committee's position that defense work should continue at Vickers alongside alternative work. Some space was devoted to outlining the alternative technologies being investigated by the committee. Less space was given to what Schofield believed was the report's main recommendation: the call for more government-backed research and development for civil marine technology. Schofield tried to compensate for this in an explanatory letter, which the *Mail* published the following day.

The next story to appear in the *Mail* also misfired for the BAEC. The committee had asked Eric Montgomery to step in as campaign secretary to help sell the report to the yard's trade unions. His task was to organize union meetings and debates at the local, district, and area levels about the recommendations made in the interim report. But the *Mail* devoted most of the piece to Montgomery's colorful political past: "Controversial former Barrow union official is set to make his comeback into the borough's public affairs after 10 years. . . . Mr Montgomery, a former Communist Party member, is currently employed as a computer operator for the Manpower Services Commission community program. . . . He has held the job for 3 months—and community program work has provided Mr Montgomery's only source of employment since he was deposed as Amalgamated Union of Engineering Workers' district secretary in 1975."[16] The *Mail* talked of "Monty" as a shipyard legend. His past, shipyard politics, and the BAEC's new report became intermingled.

"Misleading to the Point of Being Mischievous"

A row about how VSEL was ignoring and marginalizing the report ensued. Indeed, the BAEC accused the company of dismissing it out of hand. Danny Pearson accused VSEL management of a "blinkered vision of MoD work" and announced that he was lodging an official complaint. "As a founding member of the committee and a representative of 1500 members in the factory," he told the *Mail*, "I am appalled they should take this cavalier attitude."[17]

An internal VSEL directors' memorandum (figure 6.1) was then leaked to the Barrow Trades Council. That memo showed how the company planned to denigrate the BAEC's report even before reading it. The *Mail* reported this as follows: "Barrow Trades Union Council has slammed Vickers's directors for labeling the interim report on alternatives to Trident as 'misleading' and 'mischievous.' . . . Management had not seen the Schofield report before attacking it, despite a personal copy

```
        DIRECTORS' ITEMS
     8th August, 1986
     ─────────────────────────────
     THE SCHOFIELD REPORT ON ALTERNATIVE
     EMPLOYMENT TO TRIDENT

     While the Company has not seen a copy of
     the Schofield Report on Alternative
     Employment, summaries of it in the
     Evening Mail and on the radio suggest
     that it is dangerously misleading to the
     point of being mischievous.
```

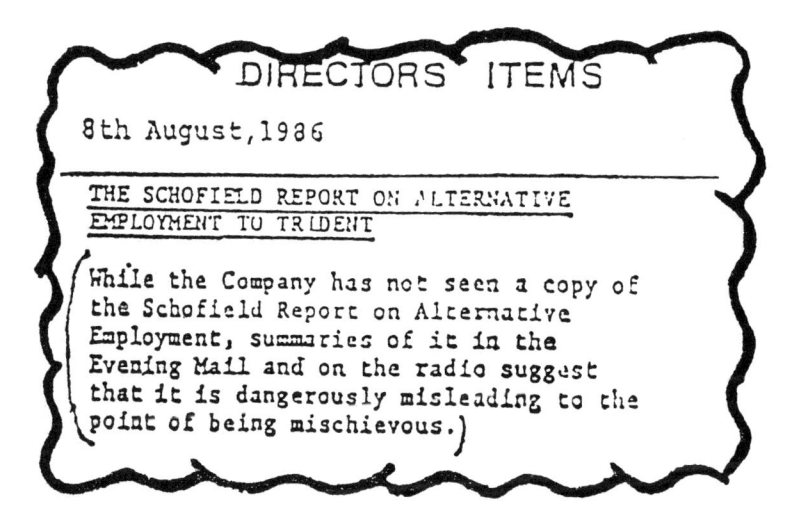

Figure 6.1
"A passage from the leaked report," *North West Evening Mail*, August 9, 1986.

being sent to Dr Rodney Leach, chief executive. . . . APEX delegate Mr Terry McSorley said: 'I cannot understand how any company which is supposed to have intelligent people at its head has not read the report and say it is misleading and mischievous.'"[18]

Management's condemnation of Employment and Security prepared the way for the company's official silence toward Oceans of Work. By the time that "technical" report was published, many uncertainties about the Trident program and about the ownership of the plant had been removed, and the report could be marginalized and dismissed as politically driven.

"Social Engineering"

Rodney Leach took the unusual step of writing directly to the *Mail* to clarify VSEL's position with respect to the interim report. Detailed study of the report, he said, had confirmed management's initial view of it as "misleading and mischievous."[19] Recruitment (for the Trident program) was proceeding steadily in Birkenhead, and selective skilled recruitment was stabilizing in Barrow.[20] It was therefore "misleading to the point of being mischievous" to claim that current defense policies were threatening the jobs of VSEL workers. Clearly referring to the upcoming general election, Leach suggested that a change in defense policy might put jobs at risk: "Statements made so far by opposition parties fall well short of the 'bankable' assurances I wish to see on employment." Steven Schofield, said Leach, had put forward alternatives "almost in the context of an attempt at worldwide social engineering, but says little about real markets for them. There is a profound difference!" The letter continued:

Those companies forming about 80 per cent of British heavy engineering which had gone into "near irretrievable decline" would doubtless also wish to diversify and get into some of the new products suggested in the report, but time and costs had to be right. VSEL's purpose designed facilities could not easily be adapted for other purposes.

It is greatly misleading to pretend otherwise to employees. The plight of merchant shipbuilding and allied marine industries the world over, and especially in the developed world does not need repetition here.

More than 80 percent of VSEL employees and many townsfolk had invested in the company on the basis of its prospects, supported by hard-nosed City investors. Perhaps Mr Schofield's talents would be better employed in starting a business to put his ideas into practice, with such financial support as he is able to raise with his own vision of the future.

This attempt to personalize the issue was, according to Schofield, a part of VSEL's strategy for not engaging with the BAEC's arguments. As Schofield pointed out in his reply (also published in the *Mail*), nowhere had Leach mentioned the BAEC or its endorsement by the Joint Staff and Manual Union Committees in the yard. No mention was made of future defense cuts, of a loss of employment resulting from the huge cost of Trident, or of the general defense trend toward fewer, more expensive vessels, aircraft, and other hardware at the expense of personnel. Management had been prepared to endorse other (non-nuclear) defense policies when it suited them—for example, during Martin O'Neill's visit, or in the VSEL Consortium's own privatization prospectus (where it gave positive emphasis to Labour's assurances of continued work on conventional submarines in the event of a new government's opting to cancel Trident). But after the buyout and the share sale, Leach appeared to be saying that these assurances were unrealistic. Schofield countered that dismissing the international recommendations about alternative marine technologies as "social engineering" only served to underline "the problems defense management have in re-orienting themselves to civilian sector production." "Perhaps," Schofield continued, "that explains the failure to capitalize on products like the Constant Speed Generator Drive, pioneered by Vickers, but successfully developed by Renk, a German company."[21] So ended the exchanges between the BAEC and VSEL in the press.

Some readers of the *Evening Mail* responded by viewing BAEC members as fairly harmless cranks. For example, two writers identifying themselves as "Nut and Bolt (Alternative Engineers)" wrote:

Inspired by reports in the *Evening Mail* of alternative strategies for Vickers we conducted our own in-depth cross-sectional poll of the population of Barrow regarding suggestions for these alternatives.

The poll took place in the snug of a well-known local hostelry, and, ignoring obviously silly suggestions, the . . . ideas for future Vickers products were:

- Grinding wheels for water mills (using non-radioactive granite). . . .
- Oars and row locks (for non-nuclear submarines). . . .
- The alternative use of the Trident shed as a permanent demo centre for CND or as a giant greenhouse to enable Barrow to come fifteenth in the Cumbria in Bloom contest. . . .[22]

At its August meeting, the BAEC heard that its researcher was to benefit from a £12,000 award as part of a grant from the Rowntree Trust to the Bradford University-based Arms Conversion Group. This would allow the research to continue beyond one year. It also meant that Schofield would be writing up the final BAEC report in Bradford, though he could continue to attend the committee's monthly meetings in Barrow.

Surveying Workers' Attitudes

The BAEC felt that it might gauge VSEL workers' responses to the promotion of alternative technologies by collaborating in a questionnaire-based survey by the Bradford group into attitudes toward defense and civil work. This study was to be carried out by Peter Southwood as a part of his doctoral research into the UK's defense industry and by Steven Schofield as a part of his research into alternatives at Barrow. The results were published by the Arms Conversion Group at Bradford in March 1987 (Southwood and Schofield 1987).

The growing interest in defense conversion, inspired by the Lucas initiative of the 1970s, was being developed by state authorities and the Office of Economic Adjustment in defense-dependent areas of the United States and by government and industrial agencies in Sweden and Germany. In the UK this interest was displayed in resolutions at national trade-union conferences and in Labour Party policies, but not by the Conservative government. Assessing the opinions of defense workers was therefore seen as an important campaign tool, and the Barrow shipyard provided a good "case study." However, it was emphasized in the final report from this study that the survey had not been a BAEC initiative.

Southwood and Schofield began by noting that, to their knowledge, there had never been "a scientific survey on the attitudes of workers and managers" in defense industries until then, although there was a large literature on the conversion of military industrial plants. Great efforts had been made to "ensure that the survey was as objective and scientific as possible." Technical, statistical, and computer analysis experts had provided assistance, and many trade unionists had given up their own time to set up the survey and administer the questionnaire. The survey's four objectives were to "test" workers' attitudes toward defense (and especially Trident) work relative to civil work, to discover the level of awareness of alternatives to military production, to discover the level of support for policies of diversification and conversion, and to discover how the special conditions prevailing in Barrow affected such attitudes. The survey was conducted by the method of systematic sampling, with respondents drawn randomly from the membership lists of four trade unions. From these lists the researchers targeted 300 individuals for interviewing. Letters of invitation were sent out by union branch secretaries, and 104 individuals (out of a workforce then totaling 12,125) were actually seen by trained interviewers. The small size of the final "sample" (which was ultimately disappointing to the researchers) was due in part to the efforts to make the survey "scientific." Each stipulation (e.g., that interviewer and interviewee not be known to each other, and that they not be members of the same trade union) reduced the potential number of participants. To protect anonymity and the confidentiality of union branch records, no "outsiders" were permitted access to membership lists. Branch secretaries were the crucial intermediary; however, for various reasons, not all branches or unions agreed to participate. Individuals who had participated in an earlier pilot study had to be eliminated because of potential "bias," and 71 interviews were "lost" as a result of confusion about the methodology. Problems also arose over balancing the input of manual and white-collar workers. In the end, the report said, the results could, at best, be regarded as "indicative of views throughout VSEL, Barrow." The survey suffered from many logistical problems, and its findings remain problematic and rather inconclusive.

Fixed-choice questionnaires, though initially seductive, often turn out to be unsatisfactory because they ignore context, because they obscure theoretical issues, and because their results are highly vulnerable to criticism. Interview on July 26, 1993, Harry Siddall said: "I hated going 'round doing the survey; it didn't seem as if we were asking the right questions."[23] To the extent that BAEC members contributed to the study, their doing so may have been ultimately counterproductive to their efforts to reach the general workforce. But the BAEC did ultimately make use of the survey's results to support its case for alternatives. Perhaps the most significant finding of the survey was the overwhelming support for a proposed pilot project involving the development of alternative technologies for the shipyard. Also, a large minority of the workers surveyed said that if given the choice they would prefer to work on civil rather than defense contracts. Very few said that they preferred defense work.[24] These "results" were considered helpful to the BAEC's case. "For the majority of the workforce," Southwood and Schofield wrote, "Trident has no special quality in terms of employment which could not be compensated by an adequate level of other defense orders."

Only about 40 percent of the workers in the sample seemed to have heard about the BAEC. A decision was made to produce a second publicity leaflet with a print run of 10,000. Eric Montgomery, who had now joined the BAEC to assist in disseminating the report, was to contact all the union branches again and send them a summary of the campaign's work. Proposals to hold a follow-up conference on industrial conversion in Barrow were again discussed at the November meeting, but it was reported that there had been no response so far from the national trade unions, which had been asked to provide speakers. The BAEC decided that Terry McSorley would write to all the relevant general secretaries. As usual, bills were paid, donations received, and conferences attended (including a fringe meeting at CND's national conference about conversion in Barrow). Schofield was planning another visit to Trident production sites in the United States.

"Technological Conservatism"

In a second short report issued by the BAEC (its precise authorship is still unclear), employment was held to be at risk as a result of mismanagement and of "technological conservatism." This report, titled Vickers and Trident—A Lesson in Management Failure, was issued during the politically sensitive period between the 1986 privatization and the 1987 general election. It contains strong criticism of management policies. It alleges that VSEL had "embarked on a clear and unequivocal campaign to ensure the continuation of the Trident program, using employment and economic arguments to disguise what is an essentially political commitment to the program's continuation." This strategy is characterized as highly dangerous and potentially disastrous for the company and the workers. Management strategy had emphasized defense exports as the best form of "diversification." But exports had been a failure. Contracts to build howitzers for India and submarines for Australia had been lost to Swedish and German companies, bids on submarine contracts for Saudi Arabia had failed, and a speculative bid on a contract to build frigates for Pakistan had disappeared without a trace. The MoD's requirements for the Type 2400 diesel-electric submarine had resulted in an oversophisticated and expensive product—more baroque technology. This meant that competitors such as the German shipbuilder HDW could manufacture submarines of similar capability that would be cheaper and more readily adaptable to technical innovation. "While Britain concentrates its design efforts on nuclear submarines," the report observed, "it is difficult to see how this will change." This export failure, along with an almost complete neglect of civil, commercial work, was the technical and social background against which Rodney Leach argued that Trident and only Trident could sustain employment at Barrow.

By appearing in the Granada TV documentary "Is Trident Good For Barrow?" alongside the Conservative local MP who was seeking re-election in the highly marginal and politically volatile constituency, and by his press statements and his letters to local and national publications, Leach reinforced the impression that the Trident program was a partisan political matter. The message was clear: only a Conservative government could guarantee Trident production. Vickers and Trident—A Lesson in Management Failure described this as "a high risk strategy" and suggested that "by trying to create the impression of dislocation through cancellation, the management may be able to arouse local fears about employment but this could be very damaging to VSEL in the national context." The BAEC maintained that, in view of the failure of the corporate strategy in defense exports and the high-risk element of a "scare campaign" to create political pressure for Trident's continuation, the workforce should be urging management to adopt a much more positive attitude toward cancellation of the project. Contingency plans for conventional defense work between the company and the MoD should be drawn up, and alternatives should be seriously considered. Otherwise, the report stated, "it is the workforce which will have to pick up the pieces."

The Last Few Meetings

The details of the BAEC's final report were hammered out at the February 1987 meeting. Its length would be about 25,000 words, and it would be published by June or July. There would be a section on product identification, a section on institutional backing, and an overview. The report would not mention Trident or Labour's non-nuclear defense policy. It would have a glossy cover and color diagrams. Because the diagrams would be important, a graphic designer would be hired. The minutes for February 5, 1987 say: "The report will be a BAEC report and nothing to do with Peace Studies. An independent report funded by this committee."

Wanting to concentrate on its original brief of investigating alternatives and protecting jobs, the BAEC declined an invitation from the Barrow council to join a committee investigating the possibility of making the district a nuclear-free zone. The BAEC replied that it was willing to provide information but did not wish to be represented on the committee.

On his return from the United States, Steven Schofield reported to the BAEC on his impressions of the US warship yards. Two US companies were competing for contracts to build Trident submarines. The largest contract so far had been won by Electric Boat, but that company was said to be looking at alternatives in case it lost future contracts. There was still very little trade-union resistance or worker-led research on alternatives in the US shipyards.

The last available BAEC records relate to meetings held on March 2 and March 31, 1987. For March 31 there is only an agenda, from which it appears that efforts were still underway to host a second national conference. BAEC members later reflected that, with the general election looming in June 1987, meetings probably were postponed and eventually the committee stopped functioning. Then came another Conservative general-election victory, and with it considerable disillusionment and exhaustion. Continuing to push for alternatives did not seem feasible. Trident had once again emerged as an election issue. Though Trident contracts were still subject to political and economic uncertainty, alternatives to Trident appeared even more unwelcome and remote.

Oceans of Work

Oceans of Work, published in August 1987, proposed eleven forms of alternative technology. These built on existing shipyard skills, but they departed from defense-oriented, MoD-dependent contracts. Thus, the report attempted to construct a new identity for the company. Oceans of Work, the long-awaited "technical study," pointed at new lines of work and advocated a different technological future.

For the United Kingdom's biggest integrated shipbuilding and engineering works, with its specific expertise in the marine environment, the report advocated development of the following:

- oceanic energy production systems
- ocean energy processing and transportation systems
- sea-based industrial systems
- open-sea mariculture and fishing systems
- ocean mining systems
- marine transportation systems for bulk, containerized, liquid, dry and gaseous cargoes
- urban mass transport systems
- riverine and sea-based public works and public utility systems
- sea-based waste disposal systems
- floating hotels, office complexes and shopping centers
- sea-based park and recreation centers.

In addition, there was a proposal for a national marine technology center—a "marine university," as Danny Pearson often described it. Such a center of excellence had always been a part of Pearson's vision for the shipyard. It would have overall responsibility for marine research and development in the UK, and it would coordinate public and private ventures. In view of the scale of the local facilities of VSFL's engineering expertise in the ocean environment, Barrow would have been a strong contender as the home for such a facility.

The language of Oceans of Work was more specific and more "technical" than that of any other BAEC publication. There were no rhetorical statements, and there was very little open criticism of VSEL management or of government policy. There were calls for support for civil marine R&D, and management received some blame for its failure to develop such civil industrial innovations as the Constant Speed Generator Drive and the Vickers Oscillating Water Column power plant; however, the report was pragmatic, positive, and largely narrative. Yet, simply because it aimed to be a "technical" report, the culmination of the researcher's findings, and a basis for arguments in favor of alternatives, Oceans of Work arguably represented the most "political" aspect of the BAEC's work. It was the fulfillment of the ambitious brief set out in that original newspaper advertisement for a researcher.

Oceans of Work was greeted by a wall of silence from VSEL. Rather than spur the BAEC on to more vigorous campaigning, its publication seemed to mark the end of the committee's activity. Things appeared to fall apart. But Steven Schofield (interviewed June 29, 1994) said he had wanted the report to be "durable": "I think we timed it very badly. . . . I wanted to make sure we'd done something very solid because I knew in a couple of years time that all those issues would be raised and that somebody in management would have to be responding to them and they did . . . at the time the whole thing was completely discredited, people just couldn't. . . . They said it was unbelievable, the committee's work wasn't credible, and yet now they're saying a lot of the things we did were right."

Oceans of Work was "launched" on November 11, 1987 at the Lisdoonie Hotel in Barrow by Bill Morris, deputy general secretary of the Transport and General Workers Union. *The Guardian* treated the launch as a minor business-page story emphasizing the UK's reluctance to fund marine R&D and the BAEC's call for a marine technology center based at the shipyard, funded by government, the European Commission and VSEL.[25]

The Campaign for Nuclear Disarmament held its own press launch of Oceans of Work in London. Schofield commented: " There was a debate with the CND about the report. They saw it as one part of a national conversion campaign, but it was essentially a trade-union document, people on the committee were under pressure to set down something directly relevant to VSEL. Oceans of Work was to be used to campaign for alternatives."

VSEL publicly ignored Oceans of Work, and, with the BAEC's initiative running out of steam, this strategy appeared to have succeeded. The report appeared when large numbers of additional workers were being recruited for the Trident contract. At that time, the company did not need to admit that concentrating on what it called its "core business" would also ultimately mean the construction of a core workforce in the region of about 4000 workers, involving the imposition of about 10,000 redundancies as Trident work diminished.

In the few letters that appeared in the local press and in the comments that were reported from meetings, there was no attempt by the company to engage with the "technical arguments." It had proved impossible for the BAEC to engage in any kind of debate with the company about products, about the content of technology. Schofield recalled: "Management were trying to suggest that all this (alternative technologies) wasn't anything to do with the trade-union movement. We kept trying to bring it back as a trade-union issue through the Trades Council." Here the role of the Confed could have been important, and parallels with the problems Lucas shop stewards encountered in getting their plan accepted by the official trade unions can be drawn. The struggle at Lucas has been seen as a struggle between old-style and new-style trade unionism (Wainwright and Elliot 1982). However, examination of both the Lucas case and the Barrow case shows that the battle was about the content and the control of technology itself.

Harry Siddall (interviewed July 26, 1993) expressed the belief that the Confed could have been crucial to fostering a wider debate about alternative technologies within the Barrow yard:

They were so wrapped up in the time . . . in trying to get more money in wages and better conditions without thinking what was going to happen in the future. It's not only management who shut their eyes, it was the Confed too, and if we could see it why couldn't they? And now they've got people going in on contract. . . . If the Confed had discussed it (Oceans of Work) properly and told all its constituent unions to study it, take it and bring it to shopfloor meetings which they could do and did with other things, then there could have been some response to it. As it is you can talk to people and they don't know what you're talking about. I talk to shop stewards that are still in the yard about Oceans of Work and they say "Oceans of Work, what's that? We haven't got any work."

Siddall and Pearson, who represented their respective trade unions on the Confed committee, knew that this body had to deal with the "bread and butter" business of wages and conditions, as they themselves had to do in their own unions. They were "old style" trade unionists; however, they were looking to the future, and they could see that sticking with huge, baroque Cold War-type technologies would lead to unemployment. Harry Siddall, in the aforementioned interview, maintained that the BAEC's repeatedly stated commitment to hybrid production and to continuation of submarine building during diversification efforts was not included in the news reports. "I always used to give a report to the joint staffs union," Siddall said, "and Dick Wade gave a report to the shop stewards' committee and we always emphasized the point that we reserved the right to build Trident, if Trident was going to be built, but nobody ever reported that. The *Mail* used to attend the Trades Council and yet never reported it." But this was a complex message. It was because of the threat of unemployment that the content of technology became the workers' business. Here is an extract from my interview with Pearson (November 6, 1991):

DP: [The Lucas Plan] was good, and some of the ideas were later taken up . . . and used by other companies.

MM: But there was a point made in that book [about Lucas] that it's very unusual and politically dangerous for workers to get involved in what they're making, that's for management.

DP: I used to start practically every negotiation I had with the company on wages or salaries or whatever. . . . I used to start, "We recognize the company's right to manage" . . . and that sets the mark and puts them at their ease you know. . . . We're not here to tell you how to run the company. . . . But at the tail end of that . . . we used to give them bloody stick if they were mismanaging the company and causing redundancies and cutbacks and whatever. . . . They have a right to manage the company, not to start mismanaging it. . . . When they start mismanaging it, they have to be accountable and first of all to the employees and their representatives.

MM: So if the product that the company is making is something that's going to destroy jobs then that's your business?

DP: Aye, that's our business.

Pearson believed that the Confed should have made the transition from "bread and butter" work (e.g., on wages) to engagement with long-term job security—in this case, with the product and the technology. Thus, the BAEC's research could have been used by the Confed in negotiations, especially if the management had wanted to impose redundancies. This was the common ground between "conventional" trade-union business and what was seen as the more radical search for alternative technologies, but it was ground that had to be very carefully prepared. Among "bread and butter convenors," said Pearson, "health and safety came quite a way behind" wages and conditions, "and you get forced into that direction both by management and the members, so you have to resist this and you have to establish that you have some priorities yourself. . . . You want the union to be identified with certain causes whether its 'free Nelson Mandela' or cleaner environment. . . . These are important things as well and you have to spend a bit of time trying to convince people. . . . We're democratic and the only way you can get to the members is to talk the them and try to convince them . . . and if you build up respect from the members as a working convenor, steward whatever, then sometimes they give you a bit of leeway, but you can't forget the bread and butter . . . because if you forget that . . . you've lost them."

Conclusion

The Barrow Alternative Employment Committee's activities were a new-style exercise in workplace democracy. The committee's questioning of the Trident program was a creative and imaginative response to the militarization of their workplace and of technology at a time when the (defense) industrial scene was flourishing locally. This was a time of, on the face of it, a full order book and full employment—"Boomtown Barrow," as it was portrayed in countless stories in the *Evening Mail*. At a time when VSEL was driving up employment to ensure Trident production, a group of workers were questioning the product and the policy. But the BAEC was not able to re-orient or "translate" VSEL, not able to turn itself into a catalyst for change, and not able to make itself an "obligatory point of passage" for successful production.

The surviving minutes of its meetings are only a partial reflection of the BAEC's activity. Operating largely on donations, the BAEC sent speakers all over the United Kingdom and beyond, informing trade-union branches, various conferences, and peace groups about its work. It received support and donations from a long list of trade unions and pressure groups. It produced two official reports, a survey, a short film, and numerous leaflets and statements. "Anyone wanting to look at alternatives here today can just get them off the shelf," Terry McSorley commented in an interview on July 31, 1992.

Of course, technologies do not just come "off the shelf." They have to be "built" both on and off the shop floor. But without a doubt the ideas about new technologies and some of the groundwork necessary to make them happen were generated by the BAEC. Some of those ideas were to emerge 5 years later in initiatives promoted by a local employment-regeneration quango,[26] Furness Enterprise, assisted by the Department of Trade and Industry, attempted to devise a form of a maritime technology center, based partly on notions of "technology transfer." These initiatives were started after huge waves of layoffs began to hit the Barrow district. And during an interview in 1993 a former BAEC activist stated that a "very senior VSEL manager" telephoned him looking for a copy of Oceans of Work when it became apparent that the Cold War had ended.[27]

Once the links in the Trident production chain were firmly secured after the 1987 general election, there was no further need for "seduction" (such as inducements to buy shares) or for "solicitation" of workers by persuasive advertising in order to strengthen the network. In interviews in 1993, current and former workers described a creeping despotism in the shipyard, hardened by the company's selective compulsory redundancy policy. (The latter will be discussed in the next chapter.) Because the network was stable, workers could have their permanent contracts terminated and then recalled if needed on short-term contracts with reduced employment benefits. Thus, a flexible "core workforce" was constructed from a pool of skilled workers. There were no "unnecessary" technological distractions and no unnecessary workers. Thus, there could be a tight, highly efficient network underpinned by high profits and record share prices.

The Barrow Alternative Employment Committee's mission had been to build and strengthen a network using research into appropriate technologies and contacts with conversion "experts," through its established contacts with workers and trade unions, and through political lobbying around industrial and political uncertainty. But a stronger configuration of harder networks militated against the BAEC's success: Trident contracts brought about "Boomtown Barrow" and the rapid growth of a dominant network. VSEL's silence toward the BAEC's initiative was a successful strategy, and its representations of the BAEC as "political" helped maintain the divide between technical content and context. Oceans of Work appeared to fail in its task; however, when large numbers of workers were sacked and when social problems increased in the local community, some of its proposals were rediscovered—too late for lost social and technical opportunities to be recovered.

IV
Closures

7
Human Redundancy: An Exercise in Disenrollment

What happens when a strong and successfully maintained sociotechnical network reaches a point where the strategic aims it embodies have been fulfilled? It may have to be altered, contracted, or even dismantled. How could the Trident submarine network be "built down" or changed? In spite of the efforts of the Barrow Alternative Employment Committee, Trident workers had been enrolled into the production of the technology. But once that production was secured and output was underway, waves of job cutting began. The widespread imposition of job cuts in the United Kingdom and in other Western industrial states—the wholesale ejection of people from sociotechnical networks—has largely been ignored by science and technology studies in general and by actor-network studies in particular.[1] Does this indicate blindness or weakness in these approaches, which otherwise have described the social-technical relationship so incisively?

Rather than telling heroic stories about the originators of the networks (as actor-network studies and other science and technology studies have tended to do), if we continue to follow the actors through production and into post-production, we arrive at the place in the life of technologies where they involve pain and dislocation. (This is not to say that earlier stages in the building of networks are free from suffering.) Network decline can be dramatic and traumatic. Workers who have been made redundant do not just disappear, even if they are ignored by technologists, many politicians, and academics. We could describe the redundancy process as one of disenrollment. How would we then describe the relationship between redundant workers and the remaining network? Has a new network been built, or has the present one been transformed? As Law (1987) says: "If the system builder is forced to attend to an actor, then that actor exists within the system."

Within Vickers, particularly when the concept of core business was being constructed, technologies (along with their social networks) whose development or continuation threatened to weaken the predominant network were marginalized. Another form of marginalization also has to take place during the final strengthening (or the dismantling) of that dominant technological network. In the Trident submarine program, large-scale recruitment gave way to selective compulsory redundancy and to the concept of creating a "core workforce" in order to retain "core skills." Thus, not only have other, distracting technologies been excised from the network; what were once major actors in that core business had also to be removed. A core workforce had to cohere around core business. In this chapter I will examine how this hegemony was achieved. The creation of a standardized representation of the workforce was integral to the method by which redundancies could be imposed and a core workforce could be shaped.

Shrinking Networks

The decline of networks, as well as their ascendancy, should be open to social studies of technology. There have been studies of failure to stabilize networks—Callon's (1986b) work on the attempt by Electricité de France to bring about development of the electric vehicle and Law and Callon's (1992) analysis of the fate of TSR2 aircraft are examples. But studies of networks' being deliberately contracted do not seem to have attracted actor-networkists. This is puzzling, since industrial decline and redundancy have been such prominent features of Western capitalist societies since the mid 1980s. A language has been assembled to convey this process: companies downsized to become "leaner, fitter and more competitive," the "product base was to be rationalized," firms restructured to divest themselves of certain interests in order to "concentrate on their core businesses," and so forth.

How should sociologists of technology find a way to describe redundant technology builders? Two possible approaches suggest themselves. One has to do with a strategy for maintaining the existing network, and one relates to the possibility of building future networks. In the first, the worker who has been made redundant is (especially in an industrially and geographically isolated community such as Barrow in Furness), a constant representation and reminder for other workers and for neighbors of the precariousness of employment within the dominant network. The former worker, by his or her absence from work, holds an almost disciplinary function for former fellow workers. By their continued presence in the community, redundant workers encourage a redoubling of diligence and compliance in the workplace in an effort to avoid the same fate. In this sense, the former worker still "belongs," or has some place, in the network. Workers who have been jettisoned from a previously tight industrial network might be described as absent intermediaries.[2] They still serve to reinforce the existing network. Thus, while the Trident network is being reduced in size, partly by the exclusion of workers, it is being tightened and strengthened by the creation of a core workforce. The company's rationale of a "core workforce" and a "core business" seeks to tie it ever closer to the Ministry of Defence and to possible future defense contracts, even though it is acknowledged that a contract as large as that for Trident will never appear again.

Another aspect of redundancy in a contracting network relates to a process of weakening the perceived importance or status of the existing network so that an altered network can appear more attractive. Reshaping the network can be done by severing some associations while strengthening others. In this way, groups that are seen to be campaigning for continued employment by arguing for the present industry to be retained or strengthened can be cast (rather patronizingly) as misguided or as behind the times. Comments made by former Defence Secretary Michael Heseltine in his capacity as president of the Board of Trade illustrate this re-framing, this drive to re-shape attitudes toward the existing network. When asked in a television interview about the impact of redundancies among prime MoD contractors such as VSEL—contractors that had been producing the major platforms for the highly sophisticated technologies of the Cold War—Heseltine said that government should not intervene to support "the skilled teams in existence doing the things they were doing yesterday, serving yesterday's markets."[3]

In 1994, when the Tyne yards of Swan Hunter Shipbuilders collapsed for lack of orders, the United Kingdom, which at one time built 80 percent of the world's unsubsidized ships, effectively abandoned major shipbuilding (Goodman and Honeyman 1988, pp. 165–166). Shipbuilding was cast as passé, along with steelmaking and coal mining. Heavy industry is often called "traditional," meaning unsophisticated or low-tech. It can be strongly argued, however, that heavy engineering and shipbuilding are both high-skill and high-tech industries. In no case would this argument be stronger than in the construction of nuclear-powered, nuclear-armed submarines.

In spite of the alleged baroqueness of the nuclear submarine fleet, the Trident shipyard could not, in 1993, have been accused of hanging onto outdated technical processes, or of aiming at a disappearing market. The four Trident submarines were technically futuristic in terms of design and component technologies.[4] (The irony is that the technical advances and design for this new class of nuclear submarine—the very ambitiousness of the project—also contribute to its "baroqueness" and its lack of industrial synergy and dynamism.) The market for Trident was generated by the Cold War; once the Cold War edifice crumbled, this market was no longer a viable network. The Cold War market was therefore a construct that, when deconstructed by history, left the Barrow shipyard workers marooned in a high-skill, high-tech industry with no clear market.

Yet Barrow had been shielded from the general decline in UK shipbuilding because of its concentration on nuclear submarines—the "gravy train," as a former Vickers director put it in an interview (conducted at Barrow) on July 26, 1993. From the Geddes Report in 1966 through the British Shipbuilders privatization in the 1980s, there had been successive attempts to differentiate the high-tech naval shipyards, and the submarine builders in particular, from the allegedly low-tech merchant vessel builders.[5] But once the market for nuclear submarines began to crumble, commentators such as Michael Heseltine sought to re-identify plants such as Barrow with that general (traditional, low-tech) shipbuilding decline, in order to present the resulting employment problems as inevitable. Heseltine's remark "Britain cannot go on supporting yesterday's markets" implies that, rather than undertaking the most ambitious submarine project in the world outside the United States, the Barrow shipyard had somehow been propped up by state aid in an ill-fated attempt to manufacture a product that belonged to a bygone industrial era.

Ironically, the BAEC forecast that the shipyard and Trident itself would become actors in one of "yesterday's markets" and that high levels of redundancy would result. However, from a financial perspective the high redundancy coincided with high profits and financial reserves. The attempt in 1986 to turn workers into shareholders could be seen as an attempt to fertilize the unproductive ground between the opposing positions of short-term profit and long-term employment. However, as was detailed in chapter 4, large institutional investors, taken together, held more shares than workers from the outset. Too many workers sold their shares too quickly for them to become financially independent of the company and independent of the economic effects of the Trident wind-down.

Thus, while the Trident network, measured by industrial production, went into decline, it was still financially strong. Share prices continued to break records, and the company became the subject of another round of takeover bids—this time by British Aerospace and GEC, each of which saw VSEL as strategically important to its business.[6] While the company shrank, it amassed a "cash mountain" of profits from Trident production, which proved attractive to bidders. Any direct relationship between the financial surplus and the redundancies was denied,[7] but the slimming down of the company by means of redundancies, closings of workshops, and dismantling of on-site manufacturing capabilities were elements of a contraction strategy to accompany the decline of work on Trident. Concurrent with a weakening of VSEL's ties with the Barrow community in the mid 1990s was a strengthening of its association with the stock market.

"Disenrollment" is used in this book as a term for the pursuit of a policy of redundancy after large-scale enrollment of Trident workers. References have been made to Trident workers as economic conscripts into the program. "Enrollment" is usually used to mean persuasion rather than coercion, but I think there is room within the concept of sociotechnical networks for some or even most of the actors to be present because they have had little choice. Again there is the issue of ambivalence—if you are an economic conscript, you may resent it and thus you may participate in the network unwillingly. For some at Barrow this ambivalence was painful and stressful. For example, the TASS convenor Danny Pearson, referring to the sinking of the *Belgrano* during the Falklands War, said "Morally we were in it up to here." There was undoubtedly a moral dilemma for him in working on Trident, and he was not alone. Because all the BAEC's members were VSEL workers, all of them were implicated in the Trident program, so they were both participating in that network and actively resisting it. They were resisting the technology from the inside. Their enrollment was in this sense only partial. Some actor-network studies have explored ambivalence. Vicky Singleton highlighted the coexistence of ambivalence and enrollment in her discussion of women's participation in cervical screening. She observed that actor-network theory tends to simplify and silence "other" (ambivalent, partially enrolled) voices because it has been concerned to account for the triumphs and failures of technologies (Singleton 1992, pp. 386–387). But if we are serious about following the actors we will have to allow room in our narrative for their contradictions. (Not too much room, of course, or our studies would never get written!)

I have discussed the privatization of the shipyard and the promotion of an employee-led buyout as a management enrollment strategy that had limited effect—workers bought shares in vast numbers, but then soon sold them and went on a lengthy all-out strike. They bought into the company's enrollment strategy, but then ditched it within weeks when circumstances changed.[8] Of course, it could be argued that ultimately the company's strategy may have been successful: just as the unions appear to have been later exhausted by the long, bitter 1988 strike, those very few workers who had retained their shares later stood to make large profits, as did company directors with substantial holdings.

The strike of 1988, a very important event in Barrow's labor history, was both the biggest and the longest strike in the UK that year. Little mention of it has been made so far in this study, partly for fear of confusing it in the narrative with the 1986 all-out strike (which followed sharp on the heels of the privatization and which was a dispute over pay and demarcation). The 1988 strike was ostensibly about the freedom of workers to choose their own holidays: workers wanted flexible arrangements, management wanted a fixed shutdown of the whole plant, claiming that this made for better planning and productivity. But many commentators agree that the issue of holiday time was the "last straw" for workers already angered by a series of disputes over the erosion of working conditions. This strike happened during the peak of Trident production. The entire Barrow community was affected by the 1988 strike, which united manual and staff workers and VSEL and non-VSEL people. A book written for the joint unions by the historian John Marshall tells a story of remarkable solidarity. Two aspects of that story seem relevant to this study. First, the strike represented "the destruction of the privatization dream of one big happy family of VSEL," and "share ownership had failed to turn the Barrow worker into a capitalist" (Marshall 1989). Second, the outcome, which ultimately amounted to a defeat for the unions, had wider political significance. There is a view widely held in Barrow—a view volunteered by every one of the workers and some of the managers who gave interviews for this study—that the strike was engineered by management. Some believe this was done so that certain existing production problems could be masked by industrial-relations problems; others believe there was a drive to exhaust and destabilize the unions before the Trident wind-down and the imposition of mass redundancies. Whatever view one takes, there is little doubt that the 1988 strike was the last sign of all-out resistance in the Barrow yard or that the dispute played a part in workers' later reluctance to take action to hold back the stream of redundancies.

Hiring and Firing

But much of the bitterness aroused by the mass job cuts in Barrow centered around the "broken promises" argument, which had to do with the way many workers say they were recruited and enrolled and with the expansion of the workforce for the Trident contracts. What was sold in the 1980s as a secure job into the next century was, within 5 years, shown to have been highly unstable and vulnerable employment. Skilled labor had been attracted to Barrow in the 1980s from distant shipyard communities that had suffered earlier closings. Later, at the height of VSEL's redundancy program, the *North West Evening Mail* ran a long feature article suggesting that the company's 1987 recruitment drive had been misleading.[9] To demonstrate this, the *Mail* reproduced an advertisement that had been placed in the *Scottish Daily Record* on January 12, 1987:

ADMIRALTY QUALIFIED WELDERS—Vacancies exist within Vickers Shipbuilding and Engineering Limited for Welders qualified to A.S.M.E. IX.—VSEL offers you real job security based on a full order book and is located in one of the most pleasant corners of England on the fringe of the Lake District National Park.—For an application form. . . .

Similarly worded ads for other categories of workers, both manual and staff, were placed in newspapers in other shipbuilding towns and in Barrow itself. The recruitment drive led to an expansion of the workforce to more than 14,000.[10] The influx of workers led to the "Boomtown Barrow" image, to rising house prices, and to a general air of prosperity. The aforementioned *Mail* article quoted an extract from a campaign speech by Conservative MP Cecil Franks: "The Conservative Party's commitment to Trident guarantees jobs for the next 20 years." Yet within 5 years it was all over for most workers.

The focus of the *Mail* article was on Jim Reilly, who said he had been told during his interview in Glasgow that his job as a shipwright in Barrow would be safe for 15 years (which, to him, represented the rest of his working life). Reilly had recently lost a job at the Greenock shipyard in Scotland. Now, 5 years later, at age 52, he was unemployed again. The move to Barrow had meant splitting up his family. Reilly took on a house mortgage in Barrow after selling his home near Greenock. Now he couldn't make his payments. Verbal promises were still binding in Scotland, he pointed out. Another welder, Jim McNeil, described how he felt he'd been "lured" to Barrow from Glasgow after answering the advertisement in the *Daily Record*. VSEL had held the interviews in Greenock to make it easier for jobless shipyard workers to attend. McNeil recalled being told that not only was there work for him but there might be apprenticeships for his son and his daughter. He to sold his home and moved to Barrow. Another worker, Paul Mills, told of answering an ad in the *Sunderland Evening Echo,* having gone 4 years without a job after the closing of the Wearside shipyards. Now, after another 4 years, having moved his family to Barrow from the North East, he was again facing unemployment. And Mills had another grievance: the criteria VSEL had used to assess him: "A man I'd never even seen before walked up to me with a yellow sheet of paper saying I am one of the 30 worst shipwrights at VSEL. Then he says I'm redundant. He says he's assessed me. He's not a shipwright himself. How can a man who doesn't know what a shipwright does and who's never seen me before assess me? I don't recognize him, and I don't recognize the bloke he's talking about in the assessment. How can they tell me I can't do my job?"

Toward Redundancy Criteria [11]

Redundancy schemes, with their system of payments, always create a very sensitive political situation for trade unions, which have to balance loss of their own power through declining working membership with the interests of individual members. Though opposition in principle to redundancy has to be the starting point, a negotiated agreement is usually seen the best possible outcome, with (to put it crudely) a "last in, first out" system reckoned to be relatively, if painfully, equitable. However, at the Barrow shipyard no negotiated agreement was reached, and job cuts were imposed.

In the summer of 1990, VSEL announced that 550 cuts in the white-collar technical staff were required, since much of the design work on Trident had been completed. The unions, led by MSF[12] senior negotiator Danny Pearson, proposed an overtime ban across the entire workforce. But the bitter experience of the 11-week all-out strike in the summer of 1988 was still fresh in workers' minds. The question put to the workers' mass meeting was: "Are you prepared to resist compulsory redundancies with action up to, and including, an overtime ban?" Although the call for this industrial action was supported on a show of hands, it was later overturned by the secret-ballot vote that was held to comply with recently introduced trade-union legislation. Early in 1991, unable to support a situation in which overtime was being worked while redundancies were being imposed, Pearson resigned from VSEL (interview with Pearson, November 6, 1991).

On November 16, 1990, after a rumor was reluctantly confirmed by management, the news broke that 1500 more workers across all departments would be shed within 12 months. The official yard unions, taken by surprise, were not able to produce a response on that day, but a member of the by-then-dormant BAEC, Bob Bolton, told the *Evening Mail*: "We told management about the need to diversify 4 years ago and it was just chucked in our face. It has been left too late."[13] Nowhere in the statement hastily prepared by management that day was there any mention of Trident. Changes in defense spending and the government's Options for Change review[14] were blamed for causing job losses in Barrow, not the declining Trident workload. On the same day, the order for the third Trident submarine was officially placed, and on the previous day a large half-yearly increase in profits had been announced. The company was careful not to link redundancies directly to the winding down of Trident work. Yet clearly recruitment specifically for the Trident program took place and huge job losses, amounting to more than half of the workforce, followed production progress through the Trident contracts.

With the announcement of 1500 job cuts, it was clear that the industrial situation was worsening and that there would be more cuts if no new orders were won by VSEL. Union officials again called for an overtime ban. Interviewed on January 30, 1995, Clive Kitchen, a former EETPU convenor,[15] recalled the attempt to persuade the Confed unions to support limited action: ". . . we never even got to the members, we were stopped in Confed, the Confed representatives, which were the shop stewards, decided against going back (to the members). We didn't want them to go on strike, we just wanted to maximize job opportunities and minimize redundancies. We didn't want to jeopardize future orders."

Agreements already existed between management and the unions whereby different shift systems could have been introduced. Kitchen believed that an overtime ban would have helped protect jobs because where two people were working 12-hour shifts management would have been forced to institute three-person 8-hour shifts. But the company drew up a selective redundancy policy and a detailed procedure for the assessment of workers by managers. The "redundancy criteria" procedure was formulated after a member of the electricians' union successfully challenged one of the first layoffs based on skill grouping before an employment tribunal.

At first the trade unions refused to discuss the redundancy criteria being formulated by management. After the details became known, different unions responded in different ways. The manual trades unions would accept nothing less than negotiations based on the "last in, first out" principle. Unions with white-collar members wanted to attempt to exert some influence over the contents of the criteria. Interviewed on December 16, 1992, Roger Henshaw of MSF, then chair of the Confederation of Shipbuilding and Engineering Unions, said:

You bear in mind that if you save one [job] somebody else goes. It's a terrible position to be in. . . . The staff group had a meeting with the company and said this criteria is rubbish, I think destroying it very effectively. We went through the criteria . . . said it's clearly absurd, that in reality you're not applying the criteria what you're doing is looking at people's sickness record . . . and the subjective judgement that the manager of the department and whether he likes somebody or doesn't like somebody . . . the criteria allowed the manager to adjust, based on subjective judgement, the number of points, to effectively select who they wanted to go. . . . And the company thanked us for our constructive comments and they haven't even bothered to come back.

According to Kitchen, EETPU was one of the unions that sought to negotiate, arguing for more limited criteria less open to personal pressures:

We didn't want to be seen to be coming to some sort of agreement that would negate our members' opportunity of going to tribunal. We altered [the EETPU's] position on that, we didn't like the criteria being developed, but we thought we should get into discussions to try to make it more objective than subjective . . . and measurable. We attempted to do that and failed, so the situation is we have no agreed position on the criteria. . . . We're free to challenge it, we're free to challenge its application.

Based on the belief that it would be impossible for a manager to conduct the assessment without being swayed by personal relationships, the union argued for simpler, "fairer" criteria that, though not ideal, would have been more "objective and measurable." "All the other stuff—how do you get on with your boss, how do you react to orders, how well do you work on your own, how well do you get on with people—all this sort of rubbish, that's subjective, it's ridiculous," said Kitchen.

The EETPU argued that the criteria should be limited to length of service, disciplinary record, absence record and sickness record; however, such limitations were not taken up, and negotiations ceased.

The unions' only effective recourse was case-by-case examination of the mechanics of redundancy—that is, of whether redundancy was applied in accordance with management's own procedure in a particular instance. For example, one tribunal case concerned the definition of the "pool" of skills and whether a group of workers could be said to be unique job holders or whether they belonged to a larger pool of people with the same or similar skills. In that case, the arguments crystallized over the issue of the unit of selection, rather than over selection itself.

The electricians' union opposed "the criteria," questioned its contents, but exploited it in tribunals in cases where it believed it could demonstrate that the "rules" had not been "correctly" applied (in management's own terms). The assessment form required managers to categorize workers and then to apportion scores against performance indicators using a point system. A ranked list would then be drawn up. The scores would then be used to select workers for redundancy. This assessment process was widely condemned by workers as unrealistic, discriminatory, and open to abuse. One senior manager, said to be conducting the assessment with enthusiasm, earned the nickname "Dr. Death." Workshops in which workers believed redundancies were imminent were called "departure lounges." Draconian images abounded. For example, during "Red Time," managers would walk through the workshops, taking note of any worker who was not actively engaged in a task.[16]

The "Job Factors" column of the redundancy assessment form had only two items that could be construed as even potentially fair and accurate: "length of continuous company service" and "conduct" (although the latter is obviously highly open to challenge). Together these two items counted for 16 points, against 52 for all the others. All the other items were highly problematic, relating to judgements about an individual worker's "knowledge" of his or her job and related jobs, skill, accuracy and quality, resourcefulness, enthusiasm, self-reliance, flexibility, and response to supervision. These were highly ambiguous, contingent, and poorly explained notions relating to individuals. Dialogue about judgments being made was impossible. The preamble to the assessment stated there would be no consultation with individuals until after the assessment had been completed and until candidates for redundancy had been identified.

The highly contestable nature of the items was masked by the tabulation format which attempted to make the procedure appear something that was, according to the preamble to the redundancy selection procedure (VSEL, Barrow, February 27, 1992), "fair, equitable and can be applied with consistency to all."

The attempt to make the criteria impersonal did not impress individual workers. A former plater, interviewed on May 24, 1992, said: "I saw a form last week from a boy who'd got the sack. I knew him. That criteria, they can make it up as they go along—they can say what they like about a boy. . . . A lot of those assessing don't know a thing about the craft they're assessing. If they don't like a boy, they'll write the right things in the right columns, they'll sort him out on the criteria." A former welder, interviewed March 3, 1992, said: "This one's ambiguous . . . the way the questions are put . . . down to the manager's judgement, whether he's a high marker or a low marker . . . we presume what they've put it forward for is to satisfy the tribunal. . . . You could be the worst disciplined person and the worst timekeeper and have the least years in, but go out for a drink with the foreman or manager. . . ."

Some workers used the tables to assess themselves, sometimes informally in small work groups, to see how the system might actually work. A crane driver/slinger, interviewed May 25, 1992, said: "Ha! You do your own assessment—how skilled you are, how long you've worked there, would you do this job, would you do that. . . . Anybody with half a brain is not going to put down that they won't do this job or that job . . . so it doesn't count. What you come down to is selective redundancy. If we agree to their criteria we're finished, it'd be an open cheque. Their criteria's very strange, they're trying to confuse people, to con people."

In the absence of a commitment by the workforce to industrial action, which would have slowed the pace of the job cuts, the unions' ability to challenge the redundancies was limited to challenging the application of the redundancy criteria before the company's internal appeals panel and in cases that were taken to an industrial tribunal. The announced purpose of the criteria was to deal with "surpluses" and to allow the company to "operate efficiently" and retain a balanced workforce. Accordingly, the company said, "the criteria will be applied to all the categories or working groups in which the surpluses exist." This created the impression that labor surpluses had occurred naturally. Redundancy was somehow inevitable, rather than something that had been brought about by industrial processes. The effect of the company's redundancy criteria was to make the selection of workers for disenrollment appear to be a detached, natural process. (One almost wants to say natural selection.) The convoluted assessment form asked a set of highly nuanced questions about an individual, and the answers were translated into numbers. The final score was then used to represent the individual's relative usefulness within the network. If you were made redundant by this method, it was somehow your own fault—you had failed to score high enough to meet the standard set for continuation in your job. Yet the numbers of people to be expelled from the network had already been set.

Conclusion

The redundancy tables were a way of observing individuals en masse, de-individualizing and codifying them, and inserting them into a system that could then be managed to bring about the desired outcome. Differences between individuals, contradictions, and ambiguities were made to disappear, but under the guise of fairness and equity. The process of getting rid of workers was made "scientific" by means of this "immutable mobile" (Latour 1987). Rose (1989) describes this as a way of making three-dimensional figures into two-dimensional ones, who are then "amenable to combination in a single visual field without variation or distortion by point of view." The individual is objectified. Rose also suggests that the inscription (e.g., the redundancy criteria) makes it possible to see the object (in this case the worker) as embodying the "properties of the theory." Here, "properties of the theory" could refer to the value judgements inherent in the criteria. The assessment form made it possible to embody those values in the actions of the worker. Thus, redundancy somehow implied a degree of individual blame or failure. Taking this point a little further, one might describe the function of the tables as circular: the individual was made to disappear into a standardized entity and then to re-appear as the cause of the problem.

Insofar as enrollment is the distribution of roles and identities within a network, the criteria also served to discipline those workers who remained in their jobs. Some would end up with positive identities, such as "good, committed self-reliant worker," while others would come out of the process with negative, "all right under surveillance" identities. Like the ambivalent participant and the absent intermediary, the worker who remains under sufferance or suspicion represents another dimension of enrollment.

"The criteria," imposed without the unions' agreement, became immutable and irresistible. It became a successful tool with which to bring about both disenrollment from the network and reinforcement of that network by creating absent intermediaries and a disciplined, "core" workforce. This did not go unchallenged. Some workers said that the criteria had been applied by people who were ignorant about them and about the content of their work and who therefore had no credentials for judging them. "The criteria" was also challenged legally in tribunals, although only within narrow constraints. Thus, while "the criteria" became immutable and irresistible, it was also partially and temporarily resisted, although workers' networks were not strong enough to disrupt its application. Just as the BAEC and its report Oceans of Work could deconstruct the Trident rationale but could not ultimately influence the content of production, the trade unions could show "the criteria" to be unfair and inconsistent but were not able to prevent its execution.

8
Softening the Facts

As I was concluding the research for this book, France was testing nuclear weapons at Mururoa in the South Pacific in the face of huge protests by Polynesian people, environmentalists, and many governments (although notably not the British). One of the effects of such tests is the further hardening of nuclear facts (facts from which weapons builders in the UK and the US were to benefit before imposition of the comprehensive test ban treaty then being advocated by President Clinton). In this volume I have attempted to soften a few of the components that are needed to build hard nuclear facts. I have borrowed this concept from Donald MacKenzie (1990, p. 420), who used it incisively in his study of missile accuracy.

In chapter 1, I asked "How do technologies succeed?" I have shown that Trident production took place within a large but simplified network that was built and held in place by "heterogeneous engineering."

The Trident production network expanded and then shrank, only to stabilize at another level. That network emerged from a heterogeneous social/technical actor world in which power relations had been more fluid and locally based. It settled into a homogeneous, tighter, and more coercive actor world, tied more closely to wider national and global actors such as the stock market. In the early 1990s, to a large extent, the money markets captured VSEL. Toward the end of my research, VSEL became embroiled in a lengthy takeover battle between GEC and British Aerospace. During that process, VSEL's share price soared. The image of the locally owned and managed company, heavily promoted in the 1986 VSEL Consortium buyout brochures, had completely dissolved.

As we have seen, a large-scale technology such as the Trident submarine consists of a disciplined coherence of people and machines into a sociotechnical network. That network had to be shaped by heterogeneous means. Examples of this shaping have included the construction of core business by the weakening of pre-existing commercial networks and the marginalization of potentially competing technologies that might have been produced by VSEL. We have seen how the identity of the company was redesigned by management's buyout drive and by the attempt to reconstruct and realign workers' identities through the share-option scheme. The attempt by some workers (the Barrow Alternative Employment Committee) to influence and reshape this emerging network/identity also had to be "engineered," and as a result the campaign for alternative production was denigrated and cast as "political" rather than "technical." And the shrinking Trident network strengthened itself by moving toward the concept of a core workforce retaining core skills by means of "disenrolling" more than half of its peak workforce.

For the Trident submarine to be successfully steered through production, what counted as "technology" and what counted as "politics" had to be kept separate. But technology and politics are parts of the same activity, and in the area of nuclear weapons systems (where politics is often overt) such a separation must be continually maintained and policed.

The "technical" recommendations put forward by the Barrow Alternative Employment Committee had to be either ignored or dismissed as "politically" motivated. Serious consideration of them would have implied acceptance of the idea that technology and politics are one. The activities of the BAEC provide a very interesting example of the maintenance of the gap between technology and politics, because the members of the committee were themselves "insiders." Crucial to their insights and to their standing in the community was the fact they were all, in different capacities, nuclear submarine workers, with local relationships and with loyalties to their shipyard and to the skills of their colleagues. As insiders they were therefore not only close to the production of submarines or submarine systems but also close to the explicit knowledge and the tacit knowledge that their positions implied. That proximity led them to be *less* certain about the technical and political feasibility and desirability of producing Trident, not more certain.

Though many of the workers may not have explicitly questioned the work they did on moral or technical grounds (in the sense that it was "just a job"), their silence did not necessarily reflect certainty that the Trident program was the only technology that could, or should, have been produced. More often, that sense of certainty was articulated by managers who, though involved with the program, were actually further removed from the activity of production. (In the 1980s and the early 1990s, top managers were increasingly recruited from other companies, and they could also return to that "outside" world.) When asked whether they would support a study of the feasibility of producing alternative technologies, most workers said yes (Southwood and Schofield 1987). It was CEO Rodney Leach who urged "all within the company to have faith and confidence in the Trident program"[1] in the context of the industrial recession that was hitting the rest of the UK's shipbuilding and engineering industries.

MacKenzie (drawing on Harry Collins's work on knowledge and replication in science) explored this variation in levels of certainty in relation to groups of workers in missile programs and proponents of such programs. Where technologists and engineers were working close to the production of knowledge about missiles, he noted a heightened awareness of technical contingency and fallibility and of other potential outcomes of technology. But this sense of contingency was not as evident among people who were *organizationally* linked to the technology rather than being its originators or its producers. This more distant group was less likely to problematize the science/technology. The sense of uncertainty acknowledged by the producers and the originators was almost as great as that of a third group, which MacKenzie described as comprising those alienated from the technology and those committed to alternatives. MacKenzie (1990, p. 372) illustrated this with a certainty trough. In a development and a critique of this certainty trough, Bijker (1995) observed that people with "high inclusion" in a technological development were more likely to have a differentiated understanding of it. Such people, according to Bijker, are exemplified by the Bakelite engineers, who knew how tricky the process of manufacturing that material could be. Those with low inclusion in the development would see the development as obdurate, but for different reasons. These two groups experience uncertainty of quite different kinds. Bijker suggests that MacKenzie's certainty trough conveys a sense that both types of uncertainty are the same. For Bijker, the quality of uncertainty changes once the actor is located outside the technological frame. Bijker's critique suggests that it is necessary to convey more about the different views of technology held by the two "wings" of uncertainty.

The Barrow trade-union activists were "insiders" in the sense that they were close to production and had a high perception of uncertainties involved in single-purpose production, in homogeneity. VSEL's managers had medium inclusion, and its directors displayed the most certainty about the Trident program. Outsiders—those alienated from the technology, such as CND groups—often had the lowest inclusion and the highest uncertainty. In this sense, Bijker's analysis is useful here: as members of the BAEC were careful to point out, they found it necessary to keep the peace movement at arm's length. Production of alternatives may have been the goal of both wings, but the methods and the approaches were different. The quality of uncertainty about the program was different in important ways.

However, some active, radical trade unionists in many parts of the UK have also been very active in the peace movement. The formation of the Trade Union Campaign for Nuclear Disarmament, out of which the BAEC grew and developed, is a testimony to the complexity of these technopolitics.

Workers' Grounds for Uncertainty

Each product created by the old, diverse Vickers involved its own set of associations, its own loyalties, and its own network. Some of these technologies were very large systems with a degree of organizational autonomy, such as the cement division; some were smaller, specialized, but applicable to a wide variety of other industries, such as pump design and production or gear grinding. Others were very small, of "cottage industry" proportions, yet were situated at the forefront of research and design, such as radome development. Whatever the activity, engineers, draftsmen, welders, and electricians had to learn about and construct that technology. Each craft group and each trade group witnessed shutdowns of production and withdrawals from (mostly commercial) markets. Production itself was therefore a precarious activity, but in addition the (perhaps unreliable and simplistic)notion of technical success, of technological "winners and losers," was being undercut by another rationale: that of Ministry of Defence-driven core business that required the abandonment of products that had hitherto been classed as winners. That the core business being promoted also relied on the continuation of the Cold War and on a certain kind of defense policy heightened this sense of uncertainty.

The experience of the builders of the Constant Speed Generator Drive illustrates the last major struggle within the company between middle-scale, heterogeneous, locally produced technology and the emergent large-scale, globally linked core business. On the CSGD project, worker control and initiative at the levels of design, engineering, and front-line management were significant. In contrast, working on Trident was described as repetitive and as isolated from the customer. Many workers, particularly in the influential TASS union, heard about the CSGD, even if they did not participate in its development. Its abandonment highlighted the social and political nature of production for a large number of workers, some close to the CSGD project and some removed from it.

Another form of uncertainty resulted from the attempts to recast workers' loyalty to the company and promote identification with the Trident project, which met with ambivalence and only partial success. The disillusionment that followed the privatization of the company was reflected in the 1986 strike and in the workers' subsequent widespread withdrawal from share ownership. Where once workers' skills had been sought after and their support (and money) had been courted for the management buyout, now an increasingly coercive regime seemed to be tightening its grip on production. Workers' adherence to the Trident project was characterized by fear of redundancy, which in turn was fueled by the "all our eggs in one basket" policy of core business. While continuing to provide jobs, Trident also reflected uncertainty in employment.

The Barrow Alternative Employment Committee's reports tried to bring all these aspects of uncertainty together. They questioned the policy of withdrawal from commercial markets. The BAEC questioned reliance on the Cold War and Trident program, opposed the privatization as against the long-term interests of Barrow, and conducted original research into alternatives.

"Boomtown Barrow" was finally laid to rest when the company's redundancy program began to bite. The gains brought by Trident in terms of protecting the town from recession had been short-lived; the losses appeared to be permanent. Talk of "diversification" was resurrected, but it quickly disappeared from management statements once VSEL's links with the money markets began to strengthen. To workers who had been made redundant and who had also sold their shares, their participation in the Trident network must have seemed (to echo then VSEL chairman David Nicholson in 1985) like a gamble in which they had lost everything.[2]

MacKenzie described the heightened perception of uncertainty that flows from proximity to the production of knowledge. But the "certainty trough" curve could also be applied to the *knowledge of production* that also springs from close experience and also gives rise to uncertainty. The "certainty trough" may help account for ambivalences among some nuclear submarine workers about their work. But it may also explain why BAEC members were such a potentially significant force. As insiders, they had knowledge of, and alliances among, the technologies they were trying to reshape. These alliances could have been crucial to the outcome of their campaign, but eventually they were destabilized by "politics." The BAEC's inside knowledge threatened to open the black box (that is, the emerging fact of production)—it threatened to soften the facticity or inevitability of the Trident program. Interestingly, workers and managers who were close to the production of knowledge and who were *not* supportive of the BAEC campaign, but who were nevertheless skeptical about the direction of technology and production inside the company, described the abandonment of the CSGD and that of the cement division as "political." That successful and profitable technical projects had been canceled for "political" reasons was a view commonly expressed by individuals in this group. The boundary between technology and politics was questioned by other actors (including managers) who were critical of company decisions, just as it was maintained by other senior managers and by politicians.

But the BAEC activists were of a different order. Though they had set out to "win the technical arguments," they were quite prepared to recognize that they were doing politics. It wasn't that they did too much technical work and not enough political work (or vice versa); it was more that their resources were few or too far distant, their allies (in the trade unions) unstable or recalcitrant, their micropolitical and macropolitical timing unfavorable, and their energies exhaustible.

We have examined a sociotechnical network and how its people and machines were enrolled. Along the way, other technologies were disenrolled; then more than half of the people in the dominant network were disenrolled. In spite of the flexibility of science and technology studies, and in spite of their potential to account for the pain brought about by shrinking networks, my modest claim is that this avenue of inquiry has been a theoretical road not taken within the field.

Other Possible Stories

As Latour says, all narratives are simplifications. This story is a constructed history and one that could have been told differently. For example, I could have "continued" different groupings of actors or types of coalitions formed between different groups. I might have explored two actor worlds configured as follows.

Actor World 1	Actor World 2
Ministry of Defence	Constant Speed Generator Drive
Core business	Civil engineers
Management	Workers
Stock Market	Barrow Alternative Employment Committee

Actor World 1 cohered because of the strengthening of associations between its constituent actors and the shedding the distractions posed by associating with those of Actor World 2. What is interesting is that the actors of AW2 did not associate successfully among themselves. While the AW2 actors had some features in common, either there was no coalition building or where coalition building was attempted (i.e., between the BAEC and workers) the alliances were not sufficient to bring about a coherent actor world.

In a different research site, it might have been possible to look at the innovation process more closely and to find examples of where technical developments were resisted but then accepted—the opposite of the CSGD experience. Turning the question around, it would then be a matter of what changes in the network facilitate revised acceptance and what significance this could have for employment and for workers' creativity. Under what conditions can workers influence the shape and the direction of the content of production? How could an alliance between "technical" content and workers' interests be built?

Disenrollment and Absent Intermediaries

When I originally formulated "disenrollment," I was thinking primarily about people, about the highly visible human redundancy in this study. But the many conversations I held with longtime workers revealed that this concept should also be applied to machines and technologies. Disenrollment of workers implies threat, coercion, redundancy, and creation of the core workforce. Disenrollment of technologies implies both marginalization of skills and marginalization of resources—the creation of the core business. While redundant workers, as absent intermediaries, signify a fate to be avoided, redundant technologies (i.e., the CSGD) also remain as absent intermediaries, as knowledge in memory.

Disenrollment appears as a negative act within a network, but it is probably seen as positive by those wishing to stabilize and strengthen networks. It seems that those skilled and long-serving workers who described their experience of redundancy were actually a *necessity* or requirement of the successful completion of the Trident program. In the process of ordering a network, there will be active intermediaries that are physically absent—once present, their trace remains. They are phantoms—phantom intermediaries, residing in the memories, practices, and skills of various actors. The extent of the waste of skill caused by the destruction of heterogeneity over the past 20 years becomes clear. So many workers carried a history within them, which they marked out by products, by technical systems they had been part of. These technologies had little to do with the Trident program, but their very absence signified its dominance.

Regrets

It was not possible to follow *all* the actors. While this research brought me into contact with a wide variety of groups, from manual workers to senior managers, there is one grouping of trade unionists which I feel is largely missing from this study. This could perhaps be called the "pro-Trident" group: those workers who opposed looking for alternatives. Lack of time prevented me from rigorous pursuit of this strand of worker opinion. But there was another, more intractable reason. When I began the research, the compulsory redundancy program had begun to be imposed. There was widespread fear and disillusionment among workers. People were less likely to declare that Trident had protected jobs, that there had in fact been a "gravy train." Some workers had changed their minds about alternatives and many who had never heard of the BAEC, were asking why engineering products had been abandoned. The "pro-Trident" group had shrunk.

While I wanted to uncover workers' perspectives on building a technology of destruction, I found that many of these workers were preoccupied with the prospect of imminent redundancy. As we have seen, boom and bust—hiring and firing—had been a regular feature of working for Vickers; little wonder that there had been no previously published workers' accounts of the industry. At the time of the study, all existing accounts were managerialist.[3] However, as is natural in the case of a high-skill workforce, some workers had become senior managers in the industry. This group continued to identify largely with the shop floor and the drawing offices—with workplace knowledge and expertise—rather than with the money markets or with "core business."

A Final Word

As I finished writing this book, the Trident submarine construction program came to an end. The remaining workers were entering a post-Trident world, with a new owner (GEC Marine)[4] and the prospect of contracts for the new class of much smaller attack submarines (the Astute Class). A core workforce will build those vessels, but the technical systems to be installed will be brought in from elsewhere. Engineering has finally lost its place to shipbuilding, remaining only as a phantom.

My starting point in this study was that technologies of destruction should not be held to be inevitable outcomes of industrial production or to be seen as stable providers of employment. Accordingly, this book is a history of Trident which is *not* the history of Trident, but of what might otherwise have been. Workers as well as sociologists open black boxes. May they continue to do so, for the sake of us all.

Appendix A
Employment Data from Lazard Brothers Sale Documents, as Reproduced in Oceans of Work

Total Numbers Employed

(September 1985)

Manual		Number
Craftsmen:	Metalworking	1,067
	Outfitting	1,424
	Engineering	1,290
Total craftsmen		3,781
Craft apprentices		1,002
Craft assistants and labourers		2,168
		6,951

Staff	Number
Management	699
Technical	2,465
Finance	109
Administration	726
Services	283
Clerical	1,077
	5,359

Total 12,310

Overall Employment

Employment statistics
(October 1985)

(Split between the two sections of the company)

VEB (Engineering)	Number	VSB (Shipbuilding)	Number
Metal using	216	Steelworkers	780
Outfit	259	Outfit	496
Engineering	805	Plant Electricians	114
		Dock Electricians	232
Craft assistants and other manual	530	Fitters	433
		Pipe Fabricators	289
Other non-craft	211	Pipe Welders	31
		Others	53
Staff	1,516	Outstation	135
		Sub Total	2,585
Apps	453	Support	1,481
Total	3,990	Apprentices	629
		Staff	3,699
		Total	8,392

Total   12,382

Notes

Introduction

1. The Borough of Barrow-in-Furness is at the southern tip of Cumbria, to the north of Morecambe Bay.

2. Previously know as Vickers, Barrow and as British Shipbuilders, the company was renamed VSEL after 1986. In 1995 it was taken over by the (British) General Electric Company (GEC). In this book, "Vickers" generally refers to the pre-1986 company and "VSEL" to the 1986–1995 company.

3. For background material on US submarine production and associated political controversies, see Tyler 1986.

4. Although this study concentrates on the production of Trident submarines in the UK, it should be stressed that "Trident" is actually the name of the US-designed-and-built missile being installed in a new class of UK-designed-and-built submarine. The missile and the submarine rely on other major sites and activities, including warhead production (at Aldermaston) and refitting and refurbishment (now to be carried out at Devonport). Hence, the term "Trident system" can be used to cover a very wide network of people, social and technical resources, and artifacts. However, "Trident" as used in Barrow, where the UK's submarines were built, almost always refers to the four-submarine construction project, and that is how the term will be used in this book.

5. Estimating the cost of Trident is such an elaborate and variable process as to become rather meaningless. To give some idea of cost, here is a range of figures from different periods in time. In 1980 official estimates of Trident 1 were between £4.5 billion and £5 billion. With the switch to D5 the cost was put at £7.5 billion at September 1981 prices. Revised estimates at September 1983 prices were £8.729 billion (House of Commons, March 13, 1984, Col. 263–264). However, Chalmers (1984, p. 54) quotes official figures for March 1984 that take into account the sterling exchange rate against the dollar and sets the cost at £9.4 billion. By 1992, when the official estimate was set at £10.5 billion, Greenpeace set the cost at £33.1 billion by including operating, refit, and support costs (David Fairhall, *Guardian*, April 29, 1992). (Here and below, *Guardian* is the well-known newspaper *The Guardian*, published in Manchester and London.)

6. For a study of the cancellation of a large-scale weapons system, see Law and Callon 1992.

7. The terms "social" and "technical" are used almost interchangeably in this study, but a fine distinction is needed in order to write about the relationship between the two. See chapter 1 for a fuller exploration.

8. I use "heterogeneous engineering" throughout this study to refer to the efforts made to control and manipulate all the social and technical elements involved in technology generally and in making any specific technology "work." For a fuller explanation see Law 1987.

9. When a delegation of trade unionists from Barrow in Furness visited the US Trident shipyards at Groton, Connecticut, to meet their opposite numbers, they were struck by the lack of political and trade-union opposition to Trident. Source: interviews with Terry McSorley (December 2, 1993) and Danny Pearson (November 6, 1991).

10. That is, technologists necessarily engaged in politics.

11. Although the "logic" of mutually assured destruction rested on "soft" targeting (that is, the destruction of major cities), that of hard-target kill rested on the perceived ability to destroy enemy missile sites.

12. For the statement about C4's adequacy for the UK, McInnes (1986) cites House of Commons 36, Q 1582.

13. The term "redundancy," common in UK (and not loaded or normative), is roughly equivalent to the US term "layoff." "Redundancy situation" is a phrase commonly used by a trade union that may be about to negotiate a orderly reduction of the workforce; hence the terms "voluntary redundancy" (taken) and "compulsory redundancy" (imposed).

14. A convenor is an individual who coordinates and administers union activities within an organization, typically by bringing shop stewards and various trade groups within a particular trade union together.

15. These theoretical approaches are discussed more fully in the next chapter.

16. In their study of the failed TSR.2, Law and Callon (1992, p. 22) acknowledge that "analytically, the fact of the failure in the present project is best seen as a methodological convenience: controversy surrounding failure tends to reveal processes that are more easily hidden in the case of successful projects and institutions."

17. A Marplan poll published in *The Guardian* on April 22, 1981 showed that 53% of those interviewed were against buying Trident and only 23% in favor. In the UK, public interest in the deal to buy Trident contrasted sharply with the United States, where there was little congressional reaction and "no reaction from the general public" (McInnes 1986, pp. 19–20). On the manipulation of public opinion to counteract the upsurge in anti-nuclear protests in the run-up to the 1983 General Election, see Ponting 1989.

18. In chapter 5 I will explore further the volatile political climate of the mid 1980s, when the Trident contracts were being negotiated.

19. Rule 5 of the Rules of Method states: "We have to be as *undecided* as the various actors we follow as to what technoscience is made of; every time an inside/outside divide is built, we should study the two sides simultaneously and make the list, no matter how long and heterogeneous, of those who do the work."

20. Foucault wrote: "Power must be analyzed as something which circulates, or rather as something which only functions in the form of a chain. It is never localized here or there, never in anybody's hands, never appropriated as a commodity or piece of wealth. Power is employed and exercised though a net-like organization. And not only do individuals circulate between its threads; they are always in the position of simultaneously undergoing and exercising this power." (Gordon 1980, p. 98)

21. Actor-network studies have tended to treat actors and groups of actors as homogeneous for the purposes of constructing narratives about them. For a critique of this tendency, see Singleton and Michael 1993.

22. The Barrow Alternative Employment Committee consisted of representatives from most of the trade and staff groups. For a profile of the Barrow plant's workforce in 1985, see appendix A.

23. Source: interview with former *North West Evening Mail* editor Keith Sutton, April 4, 1993.

24. "Marriages on the telephone," *North West Evening Mail*, July 25, 1991. The editorial states that stories about families left behind by job-seeking workers "are a distressing reminder of the human scale of the catastrophe that widespread job losses are bringing to our area."

Chapter 1

1. Callon (1986b, p. 21) writes: "More than any other kind of actor, technologists may be sometimes endowed with the capacity to construct a world, their world, to define its constituent elements, and to provide for it a time, a space and a history."

2. Examples of these literatures are listed later in this chapter.

3. Kuhn (1962, p. 138) writes: "The deprecation of historical fact is deeply, and probably functionally, ingrained in the ideology of the scientific profession, the same profession that places the highest of all values on factual details of other sorts."

4. Examples of this work include Mulkay 1979, Collins 1985, and Latour and Woolgar 1979.

5. For an outline of the social construction of technology (SCOT) approach, see Pinch and Bijker 1987 and Bijker 1992.

6. In this case the relationship between the Trident submarine and the shipyard workers and between Trident itself and other potentially competing technologies, as later chapters will explore.

7. In this way, Callon's (1986a) fascinating story of drama and treachery on the Brittany coast appears elliptical, leaving the reader plagued with unanswered questions about the context of the story.

8. In researching this book, I did not attempt to gain access to classified information; instead, I focused on finding out what could be known, what could be discussed locally, and what was visible above the iceberg's water line from historical material, open literature, and interviews.

9. Strong echoes are heard here of Robert Frost's fine poem "The Road Not Taken." Although Frost is evoking a complex and subtle dilemma with perhaps a different emphasis, his expression of the relationship between people and pathways is helpful to our understanding of choice and power in technology.

10. One of the most useful introductions to this field is Bijker, Hughes, and Pinch 1987.

11. See e.g. Hughes 1987.

12. STS heroes are legion: Diesel, Pasteur, Bakeland, Edison, Bell, Ford, Electricité de France, the Charles Stark Draper Laboratory, the Bessemer Steel Association, Vickers. To be fair, the way these heroes are followed can admit others; Latour, for example, allows Pasteur to be "an effect, a product of a set of alliances, of heterogeneous materials." See the introduction to Law 1991.

13. For more on the excision of people and machines, see Mort and Michael 1998.

14. See the introduction to Law 1994.

15. For my critique of these two classic studies of weapons systems, see Mort 1994.

16. Examples: Beynon 1973; Braverman 1974; Cooley 1980; Burawoy 1985; Noble 1984.

17. This argument is pursued more fully in Mort and Michael 1998.

18. Noble (1984, p. 33) says: "Production is what everyone wants, but the only way to get production is to get the men to work."

19. Law and Callon (1992) quote *Supply of Military Aircraft* (1955, Cmnd 9388): ". . . the failure of only one link could make a weapons system ineffective."

20. According to Tyler (1986, p. 259), the credibility of the whole system was realized at a micro level: "In a nuclear submarine, a suspicion of a defect was as good as a defect."

21. Pfaffenberger (1992) argues that the facticity of technology is located in discourse and that technology becomes coercive through myth, ritual, and classification, and that where this is resisted the result is a "technological drama."

22. According to Tyler (1986, p. 203), "people with anxiety worked harder."

23. Briefly and crudely, this is the theory that technology causes social changes, that technical artifacts have, in themselves, effects on society, thus diminishing human agency. This theory has been closely debated; see Smith 1985 and Smith and Marx 1994.

24. See, e.g., IPMS et al. n.d.

25. However, in that reshaping lies the power of the signifier; in this case it is other technologies and redundant workers, as will be explored in chapter 7. Pfaffenberger (1992, p. 300) makes, I believe, a similar point: ". . . the standardized architecture of a large-scale technological system reminds a community of its [the community's] declining autonomy."

26. One example is Latour 1996, but it is hard to resist referring to a much earlier example of a story of nonhuman agency. Bicycles and their riders exchange properties in chapter 6 of Flann O'Brien's remarkable 1967 novel *The Third Policeman*.

27. This initiative is documented in Beynon and Wainwright 1979.

28. Michael (1996, p. 60) suggests that actor-network theory is "part of a long-standing, though fragmented tradition of the theorization of power, perhaps best embodied by Foucault." See also Foucault 1974.

29. In the story of Aramis, the approach has been used to narrate the "murder" of a treasured technology. See Latour 1996.

Chapter 2

1. In 1994 and 1995, VSEL was the subject of two rival takeover bids, one by British Aerospace and one by GEC. See chapter 4.

2. Schofield (1993, p. 215) writes: "VSEL's expertise lies in the integration of a whole range of subsystems within a major defense platform: the nuclear submarine, augmented by work on conventional submarines and field howitzers. The specialist requirements of nuclear submarine production include nuclear reactor safety within a moving platform, engine silencing techniques, security of communications, and exact positioning and location facilities. It is the integration of these subsystems, which in many cases have no civilian equivalent, that provide the first major obstacle. Second, and crucially, these subsystems have to interact under extremes of conditions and performance which suggest divergences, even at the systems level, between certain kinds of highly specialized defense equipment and what might be considered as similar civil systems (in

this case merchant ships and offshore rig construction)." This draws on Schofield's Ph.D. thesis (1990). Schofield's role in the BAEC is discussed in chapters 5 and 6.

3. "Diversification" is sometimes used—less problematically, I feel—to describe industrial policies pursued in the UK after the two world wars. See e.g. pp. 301–311 of Scott 1962. It is its recent application to the Barrow situation that, I feel, can be misleading.

4. Business Review by Noel Davies, Chief Executive, in VSEL Annual Report 1993.

5. Noel Davies, quoted in "Valve design pumps up VSEL's new order hopes," *North West Evening Mail*, November 17, 1992. This policy of limited diversification was later jettisoned by the company ("Yard diversification a dead duck—Davies," *Mail*, October 14, 1994). Davies is reported as saying that the experiences of the past 4 years showed that "it was not possible for a defense contractor as committed as VSEL to diversify to compensate for the huge reductions in the defence industry."

6. Source: Sir Charles Dunphie, chairman of Vickers, quoted on p. 98 of Evans 1978. A former civil servant and public relations advisor to Harold Macmillan, Sir Harold Evans was a public-relations advisor to the Vickers board from 1966 to 1977 (when shipbuilding was nationalized). Although the book unquestionably celebrates Vickers and its status as a private company, it remains an invaluable source book as it draws on Evans's unrivaled (published) access to important individuals and internal documents.

7. Chris Barrie, "GEC fires long awaited salvo—the VSEL contest is about more than a shipyard," *Guardian*, October 29, 1994.

8. Shipbuilding profits at Barrow later climbed, partly as a result of increased naval work and partly because the Marine Installation Department was taken out of the Barrow Engineering Works and incorporated in the shipbuilding works (Evans 1978, p. 85).

9. These engines were installed in one of British Rail's most widely used locomotives, the Class 24.

10. Examples of this crucial difference in trading conditions between the defense and the commercial sectors can be found in the company's internal monitoring committee minutes 1982–1984, and which are quoted extensively in the following chapter.

11. The "official" histories to which I refer are Evans 1978, to a lesser extent Scott 1962, and certain of the company's sales brochures (particularly the VSEL/Cammell Laird sale memorandum produced by Lazard Brothers and Co. for the privatization of the shipyard in 1986).

12. Certain details of this interview have been omitted to preserve confidentiality.

13. The interviewee reported being present when a .22 rifle was fired at a slab of the material as part of a test for the FAA. "It just soaked the .22 ammunition up, didn't even break through," he said.

14. "Greg Mott was the last of the chief executives with an interest in Barrow and its future. . . . He said No—he'd been in Barrow 20 years and come up through the system, understood the works, its capabilities and range of disciplines."—former VSEL technical manager, interviewed June 30, 1994.

15. Law and Callon (1992) might call the MoD the "global network," which in this case was being strengthened at the expense of commercial (more local) networks. The difference here is that, while some of Vickers Barrow's networks and actors were local, many (e.g. the cement division) were global at the same time. Law and Callon's insights are useful here (the MoD is a common actor in both studies), but they do not quite apply to the case of Trident.

16. For accounts of such boundary work, see Gieryn 1983, 1995.

Chapter 3

1. The Submarine Machinery and Testing Establishment, the MoD's naval testing facility based within VSEL's Barrow complex, was sometimes made available for the testing of commercial equipment.

2. For example, Improvements Relating to Gears (patent 1,101,131, filed in 1965).

3. Additional design and manufacturing capacity for parallel-shaft, twin-input, and single-output gears was available at Barclay Curle Ltd. in Glasgow, which British Shipbuilders had acquired in 1978.

4. This committee is described later in the present chapter.

5. COG was sold off by Vickers in 1985.

6. Calverley, who had long design engineering experience, had joined Vickers in 1957 at its Lancaster office, and had transferred to Barrow in 1962 when the Lancaster office was closed down. Over the years he had worked on numerous civil engineering projects—cement machinery, mine winders, pumps, condensers, steam turbines, diesel engines, printing presses, soap machinery.

7. J. G. Kincaid was an engineering subcontractor in Scotland whose workshop capacity was regularly used by Vickers.

8. Walter Pringle, author of the first technical paper published about the technology, was the Engineering Technical Manager at Barclay Curle.

9. VSEL has since effectively abandoned in-house gear making.

10. "The race for the constant-speed drive," *Marine Engineers' Review*, September 1983.

11. "First sale of Vickers CSGD," *Naval Architect*, March 1984, p. E107; "A constant speed drive from Wartsila," *Naval Architect*, May 1984, p. E182; "Power booster option for Sulzer's RTA," *Naval Architect*, July-August 1984, pp. E289–E290; "Renk—first with constant speed drive at sea," *Naval Architect*, September 1984, p. E354.

12. The Technical, Administrative and Supervisory Section (TASS) was, at the time, a strong union within the shipyard. It later became part of the umbrella union Manufacturing, Science, Finance.

13. "'All eggs in one basket' blast at Vickers," *North West Evening Mail*, March 22, 1985.

Chapter 4

1. "The UK government is to sell British shipbuilders 7 warship yards," *Daily Post*, July 26, 1984.

2. "Graham Day, Chairman of British Shipbuilders has commented on the proposed privatisation of the 7 warship yards," *Financial Times*, July 27, 1984.

3. "The Declining Shipbuilding Industry," *Observer*, July 29, 1984.

4. After the 1977 nationalization, the Barrow complex was allowed to keep the name Vickers. A desire that this not be confused with the old Vickers company accounts for the use of term "Vickers Barrow."

5. "Vickers has abandoned plans to buy back its Barrow-in- Furness yard from British Shipbuilders," *Times*, September 28, 1984; "Lazards merchant bank close to completing prospectus for privatisation of British Shipbuilders' Vickers Yard," *Daily Telegraph*, January 3, 1985.

6. "Proposed Privatisation of British Shipbuilders' Warship Yards," *Financial Times*, August 3, 1984.

7. "Shipbuilding trade unionists concerned by the level of opposition to the privatisation of British Shipbuilders," *Lloyd's List*, April 24, 1984.

8. "Mass meeting of 4000 hourly paid workers at Yarrow voted not to give open support to management-led employee buy out," *Financial Times*, January 17, 1985.

9. "Bids close today on 2 British shipbuilders' warship yards," *Lloyd's List*, February 15, 1985.

10. "Demands must be satisfied before shipyard privatisation," *Lloyd's List*, May 20, 1985.

11. "Government may defer shipyard privatisation," *Mail on Sunday*, March 31, 1985.

12. "Britain's shipbuilding industry," *Guardian*, February 23, 1985.

13. "Vickers takeover boost," *North West Evening Mail*, June 13, 1985.

14. "Directors hope for management buy-out," *Financial Times*, August 13, 1985.

15. "Vickers boss says sale means freedom," *North West Evening Mail*, 4 September, 1985.

16. Lazard's chairman at the time was Sir John Nott, former Defence Secretary, under whom decisions to purchase the Trident C4 missile, and later the D5 missile, were made.

17. "Secrecy over delay in yard sale brochure," *North West Evening Mail*, May 13, 1985.

18. This point was contained in a report on the privatization prepared for the Barrow Alternative Employment Committee. The whole report has not survived and this part of it was amongst papers loaned to me by the former secretary of the BAEC. Since Norman Lamont's written answer does not appear in *Hansard* (the publication that reports daily debates in the House of Commons), it is believed to be an answer to an MP's letter. In correspondence Mr. Lamont told me that his own records did not go back as far as 1985, so he was also unable to place the quote, but he did not question its existence.

19. Vickers Shipbuilding and Engineering Limited: incorporating its wholly owned subsidiary Cammell Laird Shipbuilders Limited (sale memorandum by Lazard Brothers and Co., Ltd., October 10, 1985).

20. The only four Upholder Class vessels subsequently to be built were ordered by the Royal Navy. This order was later withdrawn, and the four submarines lay unused in Barrow after being put up for sale abroad.

21. *Lloyd's List* should not to be confused here with Lloyd's Merchant Bank, which was to handle the share option for the VSEL Consortium.

22. "Future profits of Vickers Shipbuilding will be undermined by group's merger with Cammell Laird," *Lloyd's List*, November 2, 1985.

23. "Shipyard bosses open the bidding," *North West Evening Mail*, November 11, 1985.

24. "St. James appointed to handle shipyard buyout," *Campaign*, November 15, 1985.

25. "Lowest yard buy-out stake 'about £100,'" *North West Evening Mail*, 21 November, 1985.

26. "Barrow may get slice of shipyard," *North West Evening Mail*, November 30, 1985.

27. "Yard shares not a gamble," *North West Evening Mail*, November 30, 1985.

28. "7 more days for shipyard bidders," *North West Evening Mail*, December 3, 1985.

29. "Faith backs bid for yard," *North West Evening Mail*, December 3, 1986.

30. "Shun yard shares bid—unions," *North West Evening Mail*, January 30, 1986.

31. Ibid.

32. "Subs order backs our bid for the yard," *North West Evening Mail*, January 9, 1986.

33. "Yard-town gets in on action," *North West Evening Mail*, January 23, 1986.

34. HC EDM 377. In HC debate on Royal Navy, the MP said: "On 31 March Vickers/Cammell Laird will pass into private ownership. We do not know at this stage who the bidders will be and it would not be proper for me to express a firm commitment to any one of them, but I should like to express my firm commitment to the philosophy of a management-employee consortium buy out. The whole House should welcome wider share ownership. I trust that, when the decision is made, we shall have a strong, truly independent naval defence contractor." (*Hansard*, February 6, 1986, pp. 427–428).

35. "The cash behind the yard bid," *North West Evening Mail*, February 6, 1986.

36. "Yard plea to Mrs. T on sale," *North West Evening Mail*, February 22, 1986.

37. "Consortium's pledge: 'We will talk on wages,'" *North West Evening Mail*, February 13, 1986.

38. "Minister forced to retreat over Vickers sale: bidders threat thwarts postponement decision," *Guardian*, February 21, 1986.

39. John Smith, House of Commons Private Notice Question on VSEL to Trade and Industry Minister, Peter Morrison, *Hansard*, February 24, 1986, pp. 677–678. and Lord Bruce of Donnington on HL statement on VSEL Sale, *Hansard*, February 24, 1986, pp. 849–850.

40. Cecil Franks, contribution to HC Private Notice Question on VSEL, February 24, 1986, *Hansard* 677–678. Ironically, it was VSEL that subsequently closed Cammell Laird.

41. "Shipyard sale delay averted," *North West Evening Mail*, February 21, 1986.

42. "Gerard 11, goes to No 10 for his share," *North West Evening Mail*, February 24, 1986.

43. "New lines for Vickers' pledge by Trafalgar House, *North West Evening Mail*, March 4, 1986.

44. "Yard petitions go to the government," *North West Evening Mail*, March 5, 1986.

45. "Leach wins yard deal," *North West Evening Mail*, March 7, 1986.

46. Statement by Secretary of State for Trade and Industry, Paul Channon, March 7, 1986, *Hansard*, pp. 594–595.

47. Ibid., pp. 595–596.

48. "Vickers: why the rival bid flopped," *North West Evening Mail*, March 8, 1986.

49. "Vickers gets approval for £60m buy-out," *Times*, March 8, 1986.

50. "Blue collar men urged to take up shares offer," *Engineer*, March 13, 1986, pp. 6–7.

51. "Blue collar men urged to take up shares offer," Engineer, March 13, 1986, pp. 6–7.

52. "Yard pay talks pledge 'broken,'" *North West Evening Mail*, March 11, 1986.

53. "Yard share fever grips workforce," *North West Evening Mail*, March 13, 1986.

54. "Budget Boost," *North West Evening Mail*, March 20, 1986.

55. "Yard pay: strife is threatened," *North West Evening Mail*, March 26, 1986.

56. In the UK "holiday" means any time off; it is the equivalent of "vacation" in the US.

57. "Limit set as bids swamp yard," *North West Evening Mail*, March 27, 1986.

58. "Trident deal is unique," *North West Evening Mail*, May 1, 1986.

59. "Shipyard workers vote for action," *North West Evening Mail*, May 2, 1986.

60. Ibid.

61. "Vickers staff say 'No'—strike ballot backed," *North West Evening Mail*, May 13, 1986.

62. "Vickers gets approval for £60m buy-out," *Times*, March 8, 1986.

63. "Countdown to confrontation—seeds of disruption visible months ago," *North West Evening Mail*, May 12, 1986.

64. "Countdown to confrontation," *North West Evening Mail*, May 13, 1986.

65. "Shipyard shares: High hopes," *North West Evening Mail*, July 28, 1986.

66. *North West Evening Mail*, July 30, 1986.

67. *North West Evening Mail*, July 31, 1986.

68. "£4m Bonanza for shipyard shareholders" and "Bonanza Town," *North West Evening Mail*, July 31, 1986.

69. "Maggie opens Trident hall," *North West Evening Mail*, September 1, 1986; "Vickers points way to the future—Mrs. T," *North West Evening Mail*, 4 September, 1986.

70. This is open to dispute. Analysis of the changing pattern of share holding at Companies House shows that in the first five months of possible ownership,

from April to the company's Annual Meeting in August, more than 10% of workers sold their shares. Over the following year at least a further 50% closed their accounts.

71. "Yard chief hands out warning on pay rises," *North West Evening Mail*, September 6, 1986; "Now consortium pledges airfield encore," *North West Evening Mail*, September 7, 1986.

72. "Shy shareholders ban cameras at VSEL AGM," *North West Evening Mail*, August 9, 1990.

73. Ibid.

74. They are listed in 534 pages of the VSEL Annual Returns microfilm, stored at Companies House in London.

75. Microfilm reels for 1986/7/8 and microfiches for subsequent years to 1993 for Company No 1957765 (VSEL), held at Companies House, City Road, London EC1.

76. VSEL share prices became hugely inflated, rising to more than £21 during the 1996 "auction" of the company, when GEC and British Aerospace were attempting to outbid each other.

77. "Shipyards shares hit the roof" and "Jobs axe falls in latest round of VSEL economies," *North West Evening Mail*, July 8, 1993.

78. Ibid.

79. "VSEL profit climbs to £28.8m: Cash mountain grows to £270 million," *North West Evening Mail*, November 11, 1993.

80. Multiple membership of networks and the ambivalence this entails are discussed in Singleton and Michael 1993.

81. This aspect of the campaign for workers' loyalty can now be looked at ironically in the light of the 1995 takeover struggle between two "outsiders," British Aerospace and GEC. No mention was made of local ownership during this more recent struggle, which was seen as a battle for monopoly of the UK defense industry as a whole. See Chris Barrie and Martyn Halsall, "Bid battle resumes for Barrow shipyard," *Guardian*, May 24, 1995.

Chapter 5

1. This phrase, taken from an interview with Alan Milburn MP, is explained more fully below.

2. "Baroque armaments are the offspring of a marriage between private enterprise and the state, between the capitalist dynamic of the arms manufacturers and the conservatism that tends to characterize armed forces and defense departments in peacetime . . . improvements become less and less relevant to

modern warfare, while cost and complexity become military handicaps." (Kaldor 1981, p. 5)

3. Montgomery, a former AUEW district secretary, was chief convenor at Vickers Barrow until 1976.

4. For the story of Montgomery's ejection from the yard in the 1970s, see Beynon and Wainwright 1979.

5. I.e., the producer of the first sub.

6. This process has been explored in detail by MacKenzie (1990) and Spinardi (1994).

7. Later, in 1994, workers at Rosyth and at Devonport naval dockyards, were pitted against one another for the contracts to refit the Trident submarines. This highly charged political debate was arguably the last major worker enrollment into the Trident network.

8. "'Vital' or 'suicide': The Trident debate," *North West Evening Mail*, October 23, 1984. The subject of this article is a Granada Reports TV feature of the previous night in which workers were interviewed about Trident. Alf Horne, AUEW convenor at VSEL, had called Trident workers "unwilling conscripts."

9. Booth was a former Secretary of State for Employment. Many insist that, in spite of the "Trident factor," his defeat was due mainly to changes in electoral boundaries.

10. Kaldor (1976) concluded that new sea-based technologies offered the best prospect for sustainable jobs.

11. "End the blackmail," *North West Evening Mail*, October 29, 1984. Accepting a check from CND to fund research into alternatives, Terry McSorley, APEX official and later chairperson of the BAEC, told the crowd of 22,000 that workers were "victims of a "subtle form of blackmail," because they were told there were no alternatives to the manufacture of nuclear weapons. "There are alternatives, and these are limited only by imagination and the will to implement them."

12. "Lockheed leapfrogs McDonnell," *Defense News*, July 18–24, 1994, p. 8.

13. Advisory Committee on Science and Technology, *Defence R&D: a National Resource* (HMSO, 1989), quoted in Harbor 1990.

14. "Birth of a giant—Trident no 1 takes her bow," *North West Evening Mail*, March 4, 1992; "The greatest show on earth," *North West Evening Mail*, March 5, 1992.

15. "Subs for peace, jobs plea," *North West Evening Mail*, September 4, 1983.

16. Source: interview with Harry Siddall, July 26, 1993. See also "Swords into Ploughshares: Reconversion at Barrow," *Socialist Alternatives*, July-August 1986.

17. Milburn, then Northern Trade Union CND secretary, initiated the first meetings in Kendal and organized the Barrow conference that launched the BAEC.

18. "Alternatives worker starts study," *North West Evening Mail*, April 2, 1985.

19. The panel, chaired by Terry McSorley, included CND chairperson Bruce Kent and pro-nuclear Lady Olga Maitland of Women and Families for Defence.

20. Ian Walker, "Boom-Town Britain," *Observer Magazine*, October 28, 1984.

21. When I asked Pearson about the BAEC's steering in politically dangerous waters, he said "You make us sound like Jason and the bloody Argonauts!" That off-the-cuff remark has often seemed apt.

22. "Have faith in Trident—Leach," *North West Evening Mail*, November 4, 1986.

23. "Report misleading—VSEL," *North West Evening Mail*, August 20, 1986.

24. For an account of how accusations of political bias undermine technical authority in the field of public understanding of science, see Michael 1992.

Chapter 6

1. The surviving minutes of BAEC meetings, which all took place at Barrow Labour Party offices in Hartington Street, cover only the period from July 9, 1985 to March 2, 1987. It may be that those meetings held previous to July 1985 were informal, or that they formed part of Barrow Trades Council as a subcommittee.

2. Apart from Silcocks and Wade, who could not be traced, all members readily gave interviews about their involvement in the BAEC. Other trade-union delegates who attended later meetings were also interviewed.

3. "Trident will cost jobs: Healey," *North West Evening Mail*, June 12, 1985.

4. Minutes of Barrow Alternative Employment Committee, July 9, 1985.

5. "Trident: Is this the alternative?" *North West Evening Mail*, August 15, 1985.

6. Source: minutes of BAEC, August 6, 1985.

7. "Trident study man's cash plea," *North West Evening Mail*, September 25, 1985.

8. "Trident probe man is not an amateur," *North West Evening Mail*, September 27, 1985.

9. Source: BAEC minutes, September 3, 1985.

10. "Shipyard could face big job losses," *North West Evening Mail*, October 8, 1985.

11. Ibid.

12. For a comprehensive account of earlier attempts by Vickers workers to influence production and protect jobs, see Beynon and Wainwright 1979.

13. "Vickers mustn't put all its eggs in one basket," *North West Evening Mail*, March 17, 1986.

14. "Axe for Trident wouldn't hit jobs," *North West Evening Mail*, October 10, 1985. Eight years later, other ideas about using Trident in possible "substrategic" role were posited. In this scenario Trident would be used tactically, to replace the canceled Tactical Air to Surface Missile (TASM), Colin Brown, "Scaled down Trident to replace new missile," *Independent*, April 15, 1993.

15. "Try tide power, shipyard is urged," *North West Evening Mail*, August 5, 1986.

16. "Monty joins alternatives campaign," *North West Evening Mail*, August 6, 1986. According to this story: "Monty was a fitter at Vickers for 20 years but as a union official made enemies—especially during the 1968–69 plumbers and fitters demarcation dispute, which halted the shipyard, leaving thousands idle and had to be settled by a court of inquiry. After an unsuccessful attempt to get re-elected as AUEW district secretary in 1978, Monty faded from the public scene—and has been unable to obtain a full-time job since, although he obtained a degree in history and politics." The story goes on to report that Monty, though elected by a large majority as AUEW district secretary in 1968 and re-elected unopposed in 1971, was defeated by the current officer, Frank Ward, in 1975. Montgomery then applied to get his old job back at Vickers but was told the company had no intention of employing him (against custom and practice). A fitters' overtime ban was launched in support of "Monty," but fizzled out a few weeks later. A second attempt to get industrial action in support of him was defeated, leading him to claim that Vickers management and union moderates had joined forces to drive him out of town.

17. "Yard 'cavalier' on job report," *North West Evening Mail*, August 7, 1986.

18. "Anger as yard chiefs dismiss 'alternatives'," *North West Evening Mail*, August 15, 1986.

19. "Mischievous, Mr Schofield," letter from Dr. Rodney Leach, *North West Evening Mail*, August 20, 1986.

20. Shipbuilding at Cammell Laird in Birkenhead ceased in 1993.

21. "Vickers jobs: Our fight for a secure future," letter from Steve Schofield, *North West Evening Mail*, September 9, 1986.

22. "Who needs Trident?" letter from Nut and Bolt (Alternative Engineers), *North West Evening Mail*, August 12, 1986.

23. Sidall's comments map onto sociologists' critiques of "scientific" social research, such as Cicourel 1964.

24. When asked, 79% of the sample supported a pilot project; 25% said they'd prefer civil work; 4% actively preferred defense work. These figures were also

quoted in the *North West Evening Mail* ("Trident not the mainstay says survey," March 23, 1987).

25. Martyn Halsall, "Warship option urged," *Guardian*, November 11, 1987.

26. Quango: quasi-autonomous non-governmental organization.

27. It was agreed during this interview that it would be a breach of confidence to name the individuals involved.

Chapter 7

1. The ejection of people from sociotechnical networks has been addressed in relation to the decline in research systems in Central and Eastern Europe (Balazs, Faulkner, and Schimank 1995). Their work has taken a broader approach, considering the structural factors underlying this decline, and the responses to it, documenting the nature and extent of the crisis and decline in the research system.

2. According to Callon (1991, p. 134), intermediaries are "anything passing between actors which defines the relationship between them." Intermediaries might be "scientific articles, computer software, disciplined human bodies, technical artifacts, instruments, contracts and money."

3. Michael Heseltine, interviewed by Julian O'Halloran in "The Peace Penalty: Whatever Happened to the Peace Dividend?" (BBC1 *Panorama*, March 22, 1993).

4. See the introduction for a discussion of the choice of submarines and technologies.

5. The Geddes Report recommended splitting the naval and civilian shipbuilding industries, which Geddes said were like "building lorries and racing cars together" (1966, p. 136). But the report also pointed out that Vickers Barrow and Cammell Laird Merseyside, as hybrid yards, were "outside the main groupings" (p. 91).

6. Share prices are volatile and to some extent ephemeral so it is perhaps unwise to read much into their movements. But it is worth noting that in 1994–95 the VSEL low was 980 pence rising to 1548p on 24.1.95 and 1783p on 25.5.95. In October 1994 the price jumped by 252p overnight after the announcement that a "friendly bidder" (British Aerospace) had made an offer (David Bowen, "Cash-in ahead at VSEL," *Independent on Sunday* (Business), October 2, 1994). The price was again hiked when GEC entered the competition. There is an obvious comparison, which would be made by many Barrovians, that the price had moved very far from the 1986 purchase price of 100p and from the 300p region when most workers sold their shares. (See chapter 4.)

7. "VSEL profit climbs to 28.8m: Cash mountain grows to 270 million," *North West Evening Mail*, November 11, 1993.

Notes to pp. 166–183   203

8. Callon (1986b) describes this as a kind of "betrayal."

9. "I thought I had a job for life—now I'm going to get thrown on to the streets," *North West Evening Mail*, April 1, 1992.

10. The number of VSEL employees peaked in 1989 at 16,230 workers nationally. This fell to 16,610 in 1990, then to 15,464 in 1991, then dropped to 13,028 in 1992 and to 9820 by March 31, 1993. Two months later the total was 8200. (Annual Report 1993, VSEL Consortium PLC and Subsidiaries, p. 36, and Chairman's statement, p. 2.) These are national figures, but as the vast majority (more than 90%) of VSEL employees work at Barrow, the figures reflect the redundancy program and the Trident wind-down. By May 1995, the workforce stood at around 5000 with predictions of a fall to about 4000.

11. The VSEL's redundancy assessment form and preamble became known as the "redundancy criteria" (plural) because it contained a complex and detailed set of assessments to be made. But it was also a single thing, a test, an exam, leading to a single "diagnosis." Therefore, "the criteria" will be used throughout this chapter to mean the whole formal procedure for redundancy selection.

12. The union known as MSF (for Manufacturing, Science, Finance) now incorporated TASS.

13. Maggie Mort, Liza Perks, and Chris Hill, "1,500 jobs go in new yard blitz," *North West Evening Mail*, November 16, 1990.

14. This was nicknamed "Options for Closures" by the unions.

15. When I interviewed him, Kitchen was the AEEU's officer for the north of England. Previously he had been an EETPU convenor at VSEL and a branch delegate to the BAEC.

16. At Electric Boat the individuals who did this were called "floorwalkers" (Tyler 1986, p. 115).

Chapter 8

1. "Have faith in Trident—Leach," *North West Evening Mail*, November 4, 1986.

2. "Yard shares not a gamble," *North West Evening Mail*, November 30, 1985.

3. A welcome exception is Lockett 1995. This affectionate account of the Barrow shipyard is a workers' view that goes some way toward addressing the gap in local history and recognition of workers' knowledge.

4. BAe Systems, formed in a 1999 merger of GEC and British Aerospace, now incorporates the Barrow shipyard.

Bibliography

BAEC (Barrow Alternative Employment Committee). 1986. Employment and Security.

BAEC. 1987. Oceans of Work.

BAEC. Undated. Vickers and Trident—A Lesson in Management Failure.

Balazs, Katalan, Wendy Faulkner, and Uwe Schimank. 1995. "Transformation of the research systems of post-Communist Central and Eastern Europe: An introduction." *Social Studies of Science* 25, no. 4: 613–632.

Barnes, Barry. 1995. *About Science*. Blackwell.

Barrow Technical Publications Department (Engineers). N.d. Constant Speed Generator Drive.

Beynon, Hugh. 1973. *Working for Ford*. Penguin.

Beynon, Hugh, and Hilary Wainwright. 1979. *The Workers' Report on Vickers: The Vickers Shop Stewards Combine Committee Report on Work, Wages, Rationalisation, Closure, and Rank-and-File Organisation in a Multinational Company*. Pluto.

Bijker, Wiebe. 1992. "The social construction of fluorescent lighting, or how an artifact was invented in its diffusion stage." In *Shaping Technology/Building Society*, ed. W. Bijker and J. Law. MIT Press.

Bijker, Wiebe. 1995. *Of Bicycles, Bakelites, and Bulbs: Toward a Theory of Sociotechnical Change*. MIT Press.

Bijker, Wiebe, and John Law, eds. 1992. *Shaping Technology/Building Society: Studies in Sociotechnical Change*. MIT Press.

Bijker, Wiebe, Thomas P. Hughes, and Trevor Pinch, eds. 1987. *The Social Construction of Technological Systems: New Directions in the Sociology and History of Technology*. MIT Press.

Braverman, Harry. 1974. *Labour and Monopoly Capital: The Degradation of Work in the Twentieth Century*. Monthly Review Press.

Burawoy, Michael. 1985. *The Politics of Production: Factory Regimes under Capitalism and Socialism.* Verso.

Callon, Michel. 1986a. "Some elements of a sociology of translation: Domestication of the scallops and the fishermen of St Brieuc Bay." In *Power, Action and Belief,* ed. J. Law. Routledge and Kegan Paul.

Callon, Michel. 1986b. "The sociology of an actor-network: the case of the electric vehicle." In *Mapping the Dynamics of Science and Technology,* ed. M. Callon et al. Macmillan.

Callon, Michel. 1987. "Society in the making: the study of technology as a tool for sociological analysis." In *The Social Construction of Technological Systems,* ed. W. Bijker et al. MIT Press.

Callon, Michel. 1991. "Techno-economic networks and irreversibility." In *A Sociology of Monsters*, ed. J. Law. Routledge,

Callon, Michel, John Law, and Arie Rip, eds. 1986. *Mapping the Dynamics of Science and Technology.* Macmillan.

Chalmers, Malcolm. 1984. *Trident: Britain's Independent Arms Race.* CND Publications.

Cicourel, A. V. 1964. *Method and Measurement in Society.* Free Press.

Clark, Tom. 1971. *A Century of Shipbuilding: Products of Barrow in Furness.* Dalesman Books.

Collins, Harry. 1985. *Changing Order: Replication and Induction in Scientific Practice.* Sage.

Connery Latham, Edward, ed. 1971. *The Poetry of Robert Frost.* Jonathan Cape,.

Cooley, Mike. 1980. *Architect or Bee? The Human Technology Relationship.* Hand and Brain.

Coopey, Richard, Matthew Uttley, and Graham Spinardi, eds. 1993. *Defence Science and Technology: Adjusting to Change.* Harwood Academic.

Dillon, G. M. 1983. *Dependence and Deterrence: Success and Civility in the Anglo-American Special Nuclear Relationship 1962–1982.* Gower.

Ellul, Jacques. 1964. *The Technological Society.* Vintage.

Elzen, Boelie, Bert Enserink, and Wim Smit. 1996. "Socio-technical networks: How a technology studies approach may help to solve problems related to technical change." *Social Studies of Science* 26: 95–141.

Enserink, Bert. 1993. Influencing Military Technological Innovation. WMW publication 15, Department of Philosophy and Social Sciences, University of Twente.

Evans, Harold. 1978. *Vickers against the Odds 1956–1977.* Hodder and Stoughton.

Finnegan, Philip. 1994. "Lockheed leapfrogs McDonnell." *Defense News*, July 18–24, p. 8.

Foucault, Michel. 1974. "Disciplinary power and subjection." In *Power*, ed. S. Lukes. Blackwell.

Geddes, R. M. 1966. Report of the Shipbuilding Inquiry Committee 1965–66. HMSO, Cmnd 2937.

Gieryn, Thomas F. 1983. "Boundary work and the demarcation of science from non-science." *American Sociological Review* 48: 781–795.

Gieryn, Thomas F. 1995. "Boundaries of science." In *Handbook of Science and Technology Studies*, ed. S. Jasanoff et al. Sage.

Goodman, J., and K. Honeyman. 1988. *Gainful Pursuits: The Making of Industrial Europe 1600–1914*. Edward Arnold.

Gordon, C., ed. 1980. *Michel Foucault: Power/Knowledge, Selected Interviews and Other Writings 1972–1977*. Harvester Wheatsheaf.

Harbor, Bernard. 1990. "Arms conversion and military-technological synergy." *Science and Public Policy* 17, no. 3: 194–200.

Heilbroner, Robert. 1994. "Technological determinism revisited." In *Does Technology Drive History?* ed. M. Smith and L. Marx. MIT Press.

Hicks, R. J. 1970. "Experience with compact orbital gears in service." *Institution of Mechanical Engineers Proceedings* 184, part 30.

Hirschman, Albert O. 1970. *Exit, Voice and Loyalty*. Harvard University Press.

Hughes, Thomas P. 1987. "The evolution of large technological systems." In *The Social Construction of Technological Systems*, ed. W. Bijker et al. MIT Press.

IPMS (Institution of Professionals, Managers and Specialists, Manufacturing, Science, Finance, and Transport and General Workers' Union). N.d. *The New Industrial Challenge: The Need for Defence Diversification*. College Hill Press.

Irwin, Alan. 1995. *Citizen Science: A Study of People, Expertise and Sustainable Development*. Routledge.

Kaldor, Mary. 1976. The ASW Cruiser: A Case Study of Potential for Conversion in the Shipbuilding Industry. Commissioned by Vickers National Shop Stewards' Combine Committee.

Kaldor, Mary. 1981. *The Baroque Arsenal*. Hill & Wang.

Kuhn, Thomas. 1962. *The Structure of Scientific Revolutions*. University of Chicago Press.

Latour, Bruno. 1987. *Science in Action: How to Follow Scientists and Engineers through Society*. Harvard University Press.

Latour, Bruno. 1988. *The Pasteurisation of France.* Harvard University Press.

Latour, Bruno. 1996. *Aramis, or The Love of Technology.* Harvard University Press.

Latour, Bruno, and Steve Woolgar. 1979. *Laboratory Life: The Social Construction of Scientific Facts.* Sage.

Law, John, ed. 1986. *Power, Action and Belief.* Routledge.

Law, John. 1987. "Technology and heterogeneous engineering: The case of Portuguese expansion." In *The Social Construction of Technological Systems,* ed. W. Bijker et al. MIT Press.

Law, John, ed. 1991. *A Sociology of Monsters: Essays on Power, Technology and Domination.* Routledge.

Law, John. 1994. *Organising Modernity.* Blackwell.

Law, John, and Wiebe Bijker. 1992. "Postcript: Technology, stability and social theory." In *Shaping Technology/Building Society,* ed. W. Bijker and J. Law. MIT Press.

Law, John, and Michel Callon. 1992. "The life and death of an aircraft: A network analysis of technical change." In *Shaping Technology/Building Society,* ed. W. Bijker and J. Law. MIT Press.

Lockett, Alan. 1995. *The Man Wants His Boat: Stories of Barrow Shipyard.* Trinity.

Lukes, Steven. 1974. *Power: A Radical View.* Macmillan.

MacKenzie, Donald. 1990. *Inventing Accuracy.* MIT Press.

MacKenzie, Donald. 1991. How Good a Patriot Was It? *The Independent,* December 16.

MacKenzie, Donald, and Graham Spinardi. 1995. "Tacit knowledge, weapons design, and the uninvention of nuclear weapons." *American Journal of Science* 101, no. 1: 44–99.

MacKenzie, Donald, and Judith Wajcman, eds. 1985. *The Social Shaping of Technology: How the Refrigerator Got Its Hum.* Open University Press.

Marsh, Catherine, and Colin Fraser. 1989. *Public Opinion and Nuclear Weapons.* Macmillan.

Marshall, John. 1989. *The Barrow Strike 1988.* BarrowTrade Unions.

McInnes, Colin. 1986. *Trident: The Only Option?* Brassey.

Michael, Mike. 1992. "Science-in-particular, science-in-general and self." *Science, Technology, & Human Values* 17: 313–333.

Michael, Mike. 1996. *Constructing Identities: The Social, the Nonhuman and Change.* Sage.

Misa, Thomas J. 1994. "Retrieving sociotechnical change from technological determinism." In *Does Technology Drive History?* ed. M. Smith and L. Marx. MIT Press.

Mort, Maggie. 1994. "What about the workers? Essay review." *Social Studies of Science* 24, no. 3: 596–606.

Mort, Maggie, and Mike Michael. 1998. "Human and technological redundancy: Phantom intermediaries in a nuclear submarine industry." *Social Studies of Science* 28, no. 3: nnn–nnn.

Mulkay, Michael. 1979. *Science and the Sociology of Knowledge.* Allen and Unwin.

Noble, David. 1977. *America By Design: Science, Technology and the Rise of Corporate Capitalism.* Oxford University Press.

Noble, David. 1984. *Forces of Production: A Social History of Industrial Automation.* Oxford University Press.

O'Brien, Flann. 1967. *The Third Policeman* (Flamingo, 1993).

Pfaffenberger, Bryan. 1992. "Technological dramas." *Science, Technology, & Human Values* 17, no. 3: 282–312.

Pinch, Trevor, and Wiebe Bijker. 1987. "The social construction of facts and artifacts: or how the sociology of science and the sociology of technology might benefit each other." In *The Social Construction of Technological Systems,* ed. W. Bijker et al. MIT Press.

Polanyi, Michael. 1967. *The Tacit Dimension.* Routledge and Kegan Paul.

Ponting, Clive. 1989. "Defence decision-making and public opinion: A view from the inside." In *Public Opinion and Nuclear Weapons,* ed. C. Marsh and C. Fraser. Macmillan.

Pringle, G. G. 1982. "Economic power generation at sea: The constant speed shaft driven generator." *Institute of Marine Engineers' Transactions* 94, Paper 30.

Rose, Hilary. 1983. "Hand, brain and heart: A feminist epistemology for the natural sciences." *Signs* 9, no. 1: 73–90.

Rose, Nikolas. 1989. "Individualising psychology." In *Texts of Identity,* ed. J. Shotter and K. Gergen. Sage.

Schofield, Steven. 1990. VSEL Barrow and Arms Conversion: A Case Study of a Naval Shipyard in Relation to the Theory and Practice of Arms Conversion. Ph.D. thesis, University of Bradford.

Schofield, Steven. 1993. "Defence technology, industrial structure and arms conversion." In *Defence Science and Technology,* ed. R. Coopey et al. Harwood Academic.

Schofield, Steven, and Barrow Alternative Employment Committee. 1986. *Employment and Security—Alternatives to Trident: An Interim Report.* Peace Research Report 10, School of Peace Studies, Bradford University.

Schumacher, Mary. 1987. Trident: Setting the Requirements, Contracting and Epilogue. Case Study C15-88-802.0, John F. Kennedy School of Government, Harvard University.

Schumacher, Mary. 1988a. Trident: Epilogue, 1988. Draft, case study C15-88-802.0, John F. Kennedy School of Government, Harvard University.

Schumacher, Mary. 1988b. Trident Contracting (A): Drafting the Request for Proposals. Draft, John F. Kennedy School of Government, Harvard University.

Scott, J. D. 1962. *Vickers: A History.* Weidenfield and Nicholson.

Singleton, Vicky. 1992. Science, Women and Ambivalence: An Actor-Network Analysis of the Cervical Screening Programme. Ph.D. thesis, Lancaster University.

Singleton, V., and M. Michael. 1993. "Actor-networks and ambivalence: General practitioners in the cervical screening programme." *Social Studies of Science* 23, no. 2: 227–264.

Smith, Merritt Roe, ed. 1985. *Military Enterprise and Technological Change: Perspectives on the American Experience.* MIT Press.

Smith, Merritt Roe, and Leo Marx, eds. 1994. *Does Technology Drive History? The Dilemma of Technological Determinism.* MIT Press.

Smoker, Paul. 1985. Trident Town: A Study of the Opinions of the People of Barrow. Report, Richardson Institute, University of Lancaster.

Southwood, Peter. 1985. The UK Defence Industry. Peace Research Report 8, Bradford School of Peace Studies, Bradford University.

Southwood, Peter, and Stephen Schofield. 1987. Warship Yard Workers: A Survey of Attitudes to Defence and Civilian work at VSEL, Barrow. Report, Arms Conversion Group, Bradford School of Peace Studies, Bradford University.

Spinardi, Graham. 1994. *From Polaris to Trident: The Development of US Fleet Ballistic Missile Technology.* Cambridge University Press.

Spinardi, Graham. 1997. "Aldermaston and British nuclear weapons development: Testing the 'Zuckerman Thesis.'" *Social Studies of Science* 27: 547–582.

Star, Susan Leigh. 1986. "Power, technologies and the phenomenology of conventions: On being allergic to onions." In *Power, Action and Belief,* ed. J. Law. Routledge and Kegan Paul.

Trescatheric, Bryn. 1985. *How Barrow Was Built.* Hougenai.

Tyler, Patrick. 1986. *Running Critical: The Silent War, Rickover and General Dynamics.* Harper & Row.

Vickers Shipbuilding Group Ltd. Technical Publications Department. 1977. Engineering For Quality.

VSEL Employee Consortium. 1985. Share in the Action. Booklet issued by Lloyds Merchant Bank Ltd.

VSEL Marketing and Customer Services, Barrow. N.d. Submarine Radomes. Design—Manufacture —Test. VSEL's Latest Syntactic Foam Radome G-1, Fully Tested to MOD Standards for Nuclear Submarines.

Wainwright, Hilary, and Dave Elliot. 1982. *The Lucas Plan: A New Trade Unionism in the Making?* Allison and Busby.

Webster, Andrew. 1991. *Science, Technology and Society: New Directions.* Macmillan.

Winner, Langdon. 1977. *Autonomous Technology.* MIT Press.

Winner, Langdon. 1986. *The Whale and the Reactor: A Search for Limits in an Age of High Technology.* University of Chicago Press.

Wynne, Brian. 1992. "Misunderstood misunderstanding: social identities and public uptake of science." *Public Understanding of Science* 1: 281–304.

Index

Actor-network approach, 7, 9, 13, 17, 22–29, 54, 114–115, 161, 165
Alternative plans, 30, 54, 127, 143
Alternative technologies, 113, 134, 140, 144, 148–152
Amalgamated Union of Electrical Workers, 65, 124, 144
Ambivalence, 17, 21–22, 79, 107, 165, 173, 179
APEX, 120, 124, 131
Architect or Bee? (Cooley), 30
Arms conversion, 6, 56, 120, 129, 130, 142
Arms Conversion Group, 147, 148
Arms Conversion Project, 30
Asquith, Phil, 129
Association of Professional, Executive and Computer Staff, 120, 124, 131

Barclay Curle, 70, 71
Baroque Arsenal (Kaldor), 6
Barrow Alternative Employment Committee, 7–12, 33, 74, 89, 111, 113, 116–157, 174, 179, 181
Barrow Borough Council, 135
Barrow Trades Council, 118, 119, 129, 144, 154
Beynon & Wainwright, 114, 119
Bijker, Wiebe, 28, 177, 178
Black box(ing), 2, 8, 24, 29, 76–77, 117, 180, 183
Bolton, Bob, 130, 169
Booth, Albert, 114

Bradford University School of Peace Studies, 122, 134, 137, 143, 151
British Aerospace, 165, 175
British Rail, 37
British Shipbuilders, 10, 39, 55, 56, 60–63, 70, 75, 79–81, 85, 87, 91, 134, 164
Brooke Marine, 80
Burmeister & Wain, 53, 60–61, 66

Callon, Michel, 19, 24, 115, 162
Calverley, John, 56, 63, 70
Cammell Laird, 80–82, 87, 92
Campaign for Nuclear Disarmament, 118–121, 124, 129, 132–136, 153, 178
Canadian Pacific, 115
Certainty trough, 177–180
Chalfont, Lord, 6, 53
Channon, Paul, 94
Chernobyl, 18
Clinton, Bill, 175
Coercion, 17
Cold War, 1–2, 7, 24, 111, 155, 163, 164, 178, 179
Collins, Harry, 19, 177
Commonality, 5
Compact Orbital Gears, 38, 57–77
Confederation of Shipbuilding and Engineering Unions, 82, 123–125, 131, 140, 154, 169, 170
Constant Speed Generator Drive, 10–11, 36, 47, 53–77, 134, 143, 147, 153, 178–181

Contracts
 cost-plus, 5
 first Trident, 100
 fixed-price, 6
 negotiations, 92
 process, 94
 risk of cancellation, 137, 139
Cooley, Mike, 30
Core business, 10–11, 34, 45, 73, 142, 154, 162, 182
Core workforce, 154, 157, 162, 163
Counter-network, 113
Counterforce, 4
Cruise missile, 135, 136, 140

Davies, Noel, 106
Day, Graham, 80
Defense dependence, 2, 8, 34–36, 79, 115, 116, 134, 136, 143
Department of Trade and Industry, 92
Determinism, 30
 climatic, 27
 economic, 27
 intellectual, 26
Devonshire Dock Hall, 82, 83, 138
Disenrollment, 11, 23, 161, 165, 172, 181
 of technologies, 180, 181
Diversification, 28, 33–36, 45–46, 148, 150, 169, 179

Economic conscripts, 114, 115, 165
Electrical, Electronic, Telecommunications and Plumbing Union, 101, 124, 169, 170
Electric Boat Company, 2, 119, 151
Electricité de France, 162
Elliot, Dave, 127, 142, 154
Ellul, Jacques, 19
Elzen, Boelie, 19
Employment Security—Alternatives to Trident (report), 138, 139, 144
Emslie, Rick, 88
Engineering
 political, 141, 176
 social, 146, 147
 vs. shipbuilding, 37, 39

Enrollment, 11, 22, 85, 88, 96, 106–108, 113–116, 165–167
Enserink, Bert, 19
Ethiopian famine, 116
European Community Intervention Funding, 86
Evans, Harold, 36–39, 42, 47, 60

Faith, Sheila, 89
Falklands War, 118, 165
Ferranti, 30
Finnegan, Philip, 35
Fortin, Richard, 91
Franks, Cecil, 91–93, 99, 100, 167
Fuel Homogeniser, 143
Furness Enterprise, 157

Gallagher, Les, 130
GEC, 81, 89, 133, 165, 175
GEC Marine, 183
Geddes Report, 164
General election (1987), 116, 150, 152, 157
General Electric, 87
General Dynamics, 2, 24, 113, 119
General, Municipal and Boilermakers' Trade Union, 81, 100, 124
Gooding, Robert, 5
Goodman, Jordan, 163
Greenham Common protest, 116

Hall Russell, 80
Harland & Wolff, 66
Healey, Denis, 131
Heilbroner, Robert, 112
Henshaw, Roger, 170
Heseltine, Michael, 6, 82, 163, 164
Heterogeneous engineering, 3, 17, 48, 175
Hicks Flexible Pin, 57, 60–61, 72
Hicks Ray, 57–61, 67, 71
Hirschman, Albert, 25–26
Honeyman, Katrina, 163
Hughes, Thomas, 19, 22, 76
Human-machine interaction, 29–30
Hyundai, 58, 63, 66, 68, 71

Identity, 99, 173, 176
Immutable mobile, 173
Intermediaries
  absent, 163, 173, 181–182
  phantom, 29, 182
Interpretive flexibility, 5
Irwin, Alan, 18

Kaldor, Mary, 6–7, 21, 115
Kelly, Hugh, 36, 47–48
Kincaid Engineering Works, 66
Kitchen, Clive, 169, 170
Knife and forking, 67
Knowledge, 17, 19, 43, 45, 59
  scientific, 7, 18
  tacit, 17, 43
Koos, Gary, 119
Kuhn, Thomas, 18

Labour Party, 116, 125, 130, 139, 147, 151
Lamont, Norman, 85
Latham, Bill, 99, 101, 119
Latour, Bruno, 9, 17, 22, 23, 113, 115, 173, 180
Law, John, 8, 17, 19, 22, 28, 76, 121, 161, 162
Lawson, Nigel, 98
Lazard Brothers, 85–86, 89, 93, 126, 133
Leach, Rodney, 83, 90–95, 100–104, 126, 127, 131, 144, 150, 177
Lloyds Merchant Bank, 91, 96
London Stock Exchange, 95, 96, 100, 105, 181
Lucas Aerospace Shop Stewards Alternative Plan, 8, 30, 115, 120, 127, 135, 140–142, 154, 155
Lukes, Steven, 31

MacKenzie, Donald, 2, 17, 19, 20, 23, 27, 175, 177, 179
Management buyout, 33, 83, 84, 87, 143
Manufacturing, Science, Finance (union), 44–45, 168

Marine Technology Centre, 152, 153, 157
Marshall, John, 166
McInnes, Colin, 5
McSorley, Terry, 9, 120, 122, 125–130, 134, 140, 144, 146, 150, 156
Merton, Robert, 18
Michael, Mike, 31
Milburn, Alan, 118, 120, 124
Milspecs, 46
Miners' strike, 116
Ministry of Defence, 10, 69, 70, 73, 75, 80, 87, 92, 100, 115, 116, 125, 140, 144, 163, 181
Misa, Thomas, 27
Molesworth, resistance camps at, 116
Montgomery, Eric, 6, 112, 114, 144, 149
Morris, Bill, 153
Mort, Maggie, 31
Murrell, Peter, 53, 56, 60, 63, 71
Mutually assured destruction, 4, 113

Nationalization, 33, 37–38, 54, 81
Nicholson, David, 88, 94, 179
Niigata Converter Company, 59, 62, 66–68, 71
Noble, David, 10, 20, 28

Obligatory point of passage, 28, 156
Oceans of Work (report), 8, 55, 116, 129, 138, 146, 152, 174
Office of Economic Adjustment, 148
O'Halloran, Julian, 53
O'Neill, Martin, 100, 135, 137, 147
Options for Change (government review), 169
Oscillating Water Column, 153
Overtime ban, 169

Panorama (television series), 6, 53
Patents, 56, 59, 68, 74, 126
Peace dividend, 6
Pearson, Danny, 73, 117–119, 124, 125, 129, 130, 144, 152, 154, 168, 169

Pearson, Keith, 90, 97
Peebles, Edward, 5
Pershing missile, 135, 136, 140
Petitions, 93
Polanyi, Michael, 19
Ponting, Clive, 2
Pringle, G. G, 57, 68
Privatization, 63, 79, 81, 85, 117, 125, 143, 150, 166
Profits, 11

Redshaw, Leonard, 47, 73
Redundancies, 11–12, 142, 163, 168–173
  compulsory, 92, 105, 154, 162, 168, 182
  voluntary, 90
Renault, 24
Renk, 58, 62, 68, 72, 74, 147
Resistance, by workers, 151, 165, 166
Reverse salients, 22, 76
Rheinmetall, 87
Richardson Institute, 135
Rickover, Hyman, 5
Rip, Arie, 19
Roads not taken, 20, 53
Rose, Nikolas, 173
Rowntree Trust, 147
Royal Navy, 59, 140

Schofield, Steven, 34, 122–125, 130, 131, 142–149, 153, 177
Schumacher, Mary, 2, 5–6
Science and technology studies, 4–5, 11, 18, 161
Scott, J. D., 35
Scott Lithgow, 90–92
Secret ballots, 100, 107
Shareholders, 95, 96, 104–106, 166, 179
Share in the Action venture, 87, 95
Share option scheme, 11, 79, 88, 98, 108, 143
Share prices, 103, 105
Shell International, 58, 71
Shipbuilding and Allied Industries Management Association, 82, 91

Shipbuilding industry, 82, 163
Shipbuilding Negotiating Committee, 82
Siddall, Harry, 120, 124, 130, 136, 149, 154, 155
Silcocks, Roy, 130
Singleton, Vicky, 165
Skybolt missile, 115
Slingsby Sailplanes, 38
Smit, Wim, 19
SMITE, 63–66, 69
Smoker, Paul, 136
Social construction of technology, 21
Social/technical divide, 143, 176
Southwood, Peter, 134, 148, 177
Spinardi, Graham, 2, 17, 19, 23, 26, 27
Star, Susan Leigh, 23
St James Corporate Communications, 88, 90
Strategic Arms Limitation Talks, 6
Strikes
  of 1986, 100, 107, 143, 166, 179
  of 1988, 166
Submarines
  Astute class, 183
  Belgrano, 165
  Dreadnought, 115
  Oberon class
  Ohio class, 3
  Polaris, 59, 89, 111–115
  Resolution, 112
  Type 2400 Patrol Class, 64
  Trafalgar class, 8
  Trenchant, 126
  Upholder class, 1, 86, 90
  Vanguard class, 3, 164
Sulzer engines, 37, 46, 47, 58, 68, 72
Swan Hunter, 80, 83, 163
Sweeteners, 96, 97
Synergies, 43

Technical, Administrative and Supervisory Section, 73, 119, 124, 179
Technological conservatism, 126, 150
Technological determinism, 19, 27
Technological shrinkage, 13, 28–29, 36

Technological trajectory, 27–28
Technology
  baroque, 35, 135, 150
  civil vs. defense, 48, 56, 59, 139
  content and context of, 127, 140, 157
  and politics, 26–29, 127, 140, 176
  transfer of, 157
Technopolitics, 19, 178
Thatcher, Margaret, 7, 91, 93, 103
Thomas, Douglas James, 56, 62, 65–70
Todd, Ron, 129
Trade Union Campaign for Nuclear Disarmament, 118, 119, 124
Trade Union Congress, 120
Trade unions, 10, 151, 166
Trafalgar House, 79, 81–83, 89, 91, 94, 133
Trajectory of production, 112, 113. *See also* Technological trajectory
Transmissions, 54, 57, 60–63
Transport and General Workers Union, 129, 153
Trescatheric, Bryn, 35
Trident missiles, 2–5, 113, 139
Tripoli bombing, 116
TSR2 aircraft, 162
Tyler, Patrick, 2, 24, 113

Uncertainty, 176–179. *See also* Certainty trough
US Navy, 24

Vickers, 35, 63
  cement division, 40, 178
  empire, 36–39, 60
  gearing department, 70, 73
  joint monitoring committee, 58, 63, 64, 70, 134
  mechanical engineering department, 58, 64
  shop stewards, 114
  workforce, 124, 143, 149, 167
Vickers Shipbuilding and Engineering Ltd., 6, 11, 103, 116, 132
Vosper Thorneycroft, 80, 83

VSEL Employee Consortium, 88, 89, 95, 147, 175
VSR3 radome, 43–45, 178

Wade, Dick, 130, 155
Wainwright, Hilary, 30, 127, 142, 154
Wajcman, Judy, 27
Wapping dispute, 30, 116
Ward, Frank, 124, 133
Webster, Andrew, 23, 76
Williams, Donald, 56, 62, 66
Winner, Langdon, 19, 25–28
Workability, 31
Working practices, 101–102
World War II, 118
Wynne, Brian, 18

Yarrow's, 80, 81, 90